CW01496797

Learning and Innovation of Chinese Firms

De Gruyter Studies in Innovation and Entrepreneurship

Series Editor
John Bessant

Volume 6

Learning and Innovation of Chinese Firms

Edited by
Jacky Hong and Shenxue Li

DE GRUYTER

ISBN 978-3-11-071493-7
e-ISBN (PDF) 978-3-11-071500-2
e-ISBN (EPUB) 978-3-11-071502-6
ISSN 2570-169X
e-ISSN 2570-1703

Library of Congress Control Number: 2022946481

Bibliographic information published by the Deutsche Nationalbibliothek
The Deutsche Nationalbibliothek lists this publication in the Deutsche Nationalbibliografie;
detailed bibliographic data are available on the Internet at http://dnb.dnb.de.

© 2023 Walter de Gruyter GmbH, Berlin/Boston
Typesetting: Integra Software Services Pvt. Ltd.
Printing and binding: CPI books GmbH, Leck

www.degruyter.com

Contents

Part I: **Introduction and Background**

Jacky Hong and Shenxue Li

1 Introduction

China as an Emerging Innovation Economy

China's economic growth has been phenomenal, but the country has not been widely recognized as an innovation economy until recently. There has been a wide perception of Chinese firms as reliant on imitation of technology and practices of advanced economies. Chinese firms are often portrayed as global, low-cost copycat producers, an image still deeply ingrained in popular imagination of the world's audiences (Dychtwald, 2021; Luo, Sun, & Wang, 2011), which undermines our understanding of their current innovation capacity.

Although the appropriation of foreign technologies and products has been identified as a driving force for their technological advancement and economic growth (Child & Tse, 2001), evidence suggests that discursive processes of knowledge creation have taken place in China (Bi et al., 2017). Strategies of Chinese firms have shifted from duplicative imitation to accelerated or novel innovation (Yip & McKern, 2016; Luo, Sun, & Wang, 2011; Williamson & Yin, 2014). Chinese firms have leveraged technology and knowledge acquired along the path of "Go Global" to bolster their innovation capabilities at home which have been adapted to better meet domestic needs (Lyles et al., 2022), a systematic process coined "re-innovation" (*zaichuangxin*) (Lu, 2013).

This shift of focus has been fueled by a series of innovation-conducive measures and policies initiated by the Chinese government (Liu et al., 2011). Over the past few decades, the government has laid out clearly defined plans and strategies for developing an innovation-oriented economy. Some notable examples are the "Medium- to Long-Term Plan for Science and Technology Development (2006–2020)" (MLP) (OECD, 2008), Made in China 2025 (PRC State Council, 2015), and the "Bring In" and "Go Global" strategies (Lyles et al., 2022), which focus on building indigenous innovation capability. The innovation potential of Chinese firms was further enhanced by state support for both basic and applied research. Sizable resources have been channelled to scientists and entrepreneurs to spur innovation through large state-sponsored initiatives such as the "State High-Tech R&D Program" (863 Program) initiated in the 1980s and the "National Program on Key Basic Research Project" (973 Program) instigated in late 1990s (Liu et al., 2011).

Moreover, significant efforts have been made to develop a higher education system with world-class research capabilities through various projects such as the 211 Project and the 985 Project initiated in the 1990s (Li, 2004) and the Double First-Class Plan designed much later (Ministry of Education of the PRC, 2017). These projects have significantly improved Chinese universities' research strength and global standing which helped train a new generation of highly skilled workforce.

https://doi.org/10.1515/9783110715002-001

Furthermore, the Chinese state capitalist system connects its private sector to state-owned enterprises (SOEs) through multiple levels of channels (Cheng, Hua, & Tan, 2019), thereby putting pressure on traditional SOEs to perform on the global stage alongside their new and innovative private competitors. An exemplar of such interconnectivity is the Government Guidance Funds (GGFs), a unique type of public-private investment funds designed to directly and indirectly invest in companies and further the state's strategic industrial policy goals mostly related to emerging and high-tech technologies (Luong, Arnold, & Murphy, 2021). With connections between the state and private sectors, both SOEs and some private firms (e.g., Huawei and Lenovo) have grown exponentially competing directly with MNCs of advanced economies.

The push for innovation has made China the 12th most innovative country in the world and the top innovative country of the upper middle-income group (Dutta et al., 2021). The rapid pace of innovation in China rejects the line of Western thinking that innovation occurs favourably in a democratic, market economy, particularly in the private sector (Hannas & Tatlow, 2020). Chinese learning is rather unconventional (Lyles, Li, & Yan, 2014). Its innovation speed does not match other innovation patterns observed globally (Dychtwald, 2021). The focus on technology transfer from advanced economies to China by policy makers in the West is rather myopic, undermining the need for an understanding of the full spectrum of innovative activities and strategies in China, which could negatively affect the capacity of Western rivals to effectively compete with China (Weinstein, 2022).

Below we explain more about the "Bring In" and "Go Global" initiatives which have had far-reaching effects on Chinese firms' learning and innovation capabilities (Lyles et al., 2022).

Learning and Innovation of Chinese Firms Through Internationalization

Most countries have made policy efforts to attract inward foreign investment, but few have set clear strategies aimed at promoting outward foreign investment (UNCTAD, 2018). China has been an exceptional case such that its government policies attach significant weight to both inward and outward foreign investment (MOFCOM, 2017). China initiated its economic reforms in late 1970s and, since then, the government has set forth "Bring In" policies and incentives (e.g., tax incentives and the establishment of special economic zones) to attract inward foreign investment (Zhou, Delios, & Yang, 2002). This was followed by its "Go Global" policy proposed in the 1990s to support Chinese firms' participation in international markets.

The rationale for "Bring in" and "Go Global" policies is clear. Both policies are seen as crucial in supporting Chinese economic sectors to move up the innovation value chain (Lyles et al., 2022). International economic research has long established models of endogenous innovation-driven growth, emphasizing the role of foreign investment as an important channel for technological knowledge flows across national borders (Findly, 1978; Grossman & Helpman, 1991; Rivera-Batiz & Romer, 1991). Inward foreign investment has been perceived as an important source of technology and knowledge spillovers by offering host market opportunities, particularly those of developing countries (Blomström & Kokko, 1998; Cantwell, 2009; Caves, 1974). Inward foreign investment has played an important role in industrial modernization and rapid and sustained economic development of China. Between 1978 and 2017, China's GDP grew at an average rate exceeding 9 per cent annually, compared with a growth rate of 2.9 per cent for the global economy (EY, 2018). Inward FDI to China reached USD 149 billion in 2020, making the country the world's second-largest FDI recipient (UNCTAD, 2021). It is well understood that China has relied on this policy to learn from foreign investors and develop its key industries, though foreign firms are still not allowed to invest in a number of sectors.

The "Go Global" strategy, while catering for the needs of the Chinese government (Child & Rodrigues, 2005), has provided a platform for those Chinese enterprises with comparative advantages to gain access to advanced technology, managerial capabilities and new markets, and to enhance their brand recognition and competitiveness in international markets (MOFCOM, 2017). The growing volume of research on emerging market multinational corporations (EM MNCs) emphasizes the role of outward foreign investment as a springboard to aggressively acquire critical resources and capabilities needed to innovate and establish their competitive positions at home and globally (Deng, 2009; Gaur, Ma, & Ding, 2018; Luo & Bu, 2018; Luo & Tung, 2007; 2018; Rui & Yip, 2008). Buckley and colleagues (2007) observed that Chinese investors often choose to enter foreign markets with rich technological endowments. The effects of Chinese outward investment on home market productivity are often contingent upon the level of technology gap between home and host countries (Li et al, 2016). A wider gap suggests the potential for significant productivity spill over effects.

While China continues its focus on "Bring In", the government has attached more importance to "Go Global" in recent years by providing financial support and incentives. As a result, Chinese outbound direct investment (ODI) had reached to over US$1.9 trillion by the end of 2018, making it one of the world's largest foreign investors, trailing just behind the US and Japan (OECD, 2020). There was also a noticeable improvement in investment quality, that is, an expansion of investment focus from resources and raw materials to strategic acquisitions in a range of services (e.g., finance and health care) and manufacturing sectors that bring complementary benefits to Chinese firms' core businesses (EY, 2018; Luo, Xue, & Han, 2010).

Clearly, both "Bring In" and "Go Global" policies have yielded successful results. The exponential growth in FDI inflows to China has attracted tremendous research interest in purposeful knowledge transfer from foreign investors to their operations in China (Li, Easterby-Smith, & Hong, 2019; Liao & Yu, 2012; Wang, Tong, & Koh, 2004) and knowledge spillover effects – knowledge created by a foreign investor but unintendedly used by an indigenous Chinese firm for which the latter does not (fully) compensate the former (Javorcik, 2004) – on productivity (Björkman & Kock, 1995; Buckley, Clegg, & Wang, 2007; Liu, 2001; Wei & Liu, 2006; Zhou, Li, & Tse, 2002).

Meanwhile, the rapid growth of Chinese outward FDI has attracted significant research attention on the effect of OFDI on domestic innovation and reverse knowledge transfer from Chinese overseas subsidiaries to their domestic operations at home (Lyles et al., 2022). For example, Li and colleagues (2016) report that OFDI from Chinese firms has a positive impact on their home-based innovation performance moderated by three factors – competition intensity of the local market, absorptive capacity, and IFDI. The effects of Chinese OFDI on home market productivity are often contingent upon the level of the technology gap between home and host countries (Li et al, 2016). When the host country is more technologically advanced than the home country, a wider gap suggests the potential for significant productivity spillover effects on the home country.

Overview of Contributions to this Volume

In view of these recent developments, the present edited volume attempts to address some unresolved but critical issues underpinning the learning and innovation processes of Chinese firms. The chapters of this edited volume will further examine how Chinese firms have taken advantage of strong Chinese policy support to learn and innovate in the past few decades. Drawing on diverse conceptual perspectives and methodological approaches, each contributor focuses on a specific learning and innovation-related challenge encountered by Chinese firms along the paths of 'Bring In' to 'Go Global' (Lyles et al., 2022). There are nine chapters organized into four main parts (see Table 1.1).

Following our introduction chapter in Part I, there are two chapters on 'Knowledge Transfer and Managerial Cognition' in Part II. Through a qualitative case study of Haier, Sun investigates how their foreign subsidiaries located in emerging markets can overcome their resistance for sharing idiosyncratic and location-specific knowledge in host countries with the headquarters. By adopting both personal coordination mechanisms and electronic coordination mechanisms, the paradox between the mandate of global knowledge integration and subsidiary autonomy can be mitigated during reverse knowledge transfer. In chapter three, Zhang and Kong undertake an ethnography in Quanzhou city, China, aiming to explore the various patterns of managerial cognition formed and manifested in Chinese firms during transition and related impacts on learning and innovation. Based on in-depth interviews with 50

Table 1.1: Summary of chapters and their contributions.

Authors	Methodology	Level of analysis	Subject of study	Research question(s)	Key insights and contributions
Sun	Case study	Firm level	Subsidiaries	– How do subsidiaries of Chinese MNCs reverse transfer their knowledge from developing countries in conjunction with different coordination mechanisms to HQ?	– A high degree of subsidiary autonomy appears as a hindrance on reverse knowledge transfer to HQ; – Coordination mechanisms moderate the negative relationship between subsidiary autonomy and reverse knowledge transfer; – While personal coordination mechanisms facilitate subsidiaries' technology transfer as well as integration with HQ, electronic coordination mechanisms enhance the efficient communication and effective management of overseas subsidiaries.
Zhang & Kong	Ethnography	Individual level	Managers	– How the managers' cognition are formed and manifested in their pattern of behaviors in Chinese firms?	– Managers' cognition about transformation is evolving organically as new sources of knowledge emerge in local business contexts; – The entrepreneurs shaped by their cognitive framework tend to focus on immediate profit seeking, and transformation and innovation that do not lead to a clear and immediate profit making are not preferred; – "Self-fulfilling prophecy" and the difficulties in breaking away from the various elements inherent in their existing cognitive framework continue to shape and constrain the entrepreneurs' risk-taking and innovative behaviors in China.

(continued)

Table 1.1 (continued)

Authors	Methodology	Level of analysis	Subject of study	Research question(s)	Key insights and contributions
Zhou, Hong & Snell	Historical case study	Institutional level	Entrepreneurial ecosystems	– How entrepreneurial ecosystems affect the innovation process in China?	– Top three entrepreneurial ecosystems in China are driven by different cultural, social and material attributes to facilitate their evolution; – Various configuration of an entrepreneurial ecosystem affect the invention and commercialization activities.
Cao	Conceptual	Firm level	Born-digitals	– How would the iteration cycle affect the internationalization process of Chinese digital firms? – What is the difference between traditional internationalization and digital internationalization?	– The iteration cycle of traditional process of internationalization may undermine the learning speed of Chinese digital firms. – A new perspective is thus provided to understand the mechanisms of rapid internationalization of Chinese digital firms.
Mac & Evangelista	Survey	Firm level	Exporters	– What are the relationship between corporate entrepreneurship and organizational learning as well as their effect on export performance among Chinese exporters?	– Organizational learning styles and knowledge integration adopted by Chinese exporters affect their firm performance; – Chinese exporters are able to utilize appropriate approaches to learning as well as having a proper knowledge management system to transform learnt knowledge into favorable outcomes. – Learning commitment and learning style are significantly related to ICE which in turn leads to export performance.

Rowley & Redding	Conceptual	Micro and macro-level	Institutional environment	– How organizational learning and innovation can help China restructure its economy from a low cost to a high value model?	– There is the need for all organisations to prepare for the growing complexity attached to a competitive modern economy; – Innovation is intrinsically linked to learning and knowledge; – Complexity itself requires responses of greater empowerment within a society.
Cuervo & Crestejo	Case study	Firm level	Integrated resorts	– How can diversification be underpinned by learning and innovation in the hospitality and tourism industry in Macau?	– The six integrated resort operators in Macau pursued distinctive modes of innovation over the past two decades. – Depending on their emphasis on learning and innovation, some are more sustainable while others are vulnerable.

managers and participant observation, the findings suggest that the managers' cognitive framework about innovation is evolving organically and continuously shaped by their deep-seated cultural values and embedded assumptions.

Part III, 'Entrepreneurship, Learning and Innovation', consists of three chapters, each of which addresses the learning and innovation related issues and challenges from a system and firm-level perspective. The chapter by Zhou, Hong and Snell distinguishes three types of entrepreneurial ecosystems in China and their cultural, social and material attributes and corresponding impacts on firm innovation. Their comparative analysis of various configuration of ecosystem attributes serves to provide a holistic perspective to examine entrepreneurship and innovation for Chinese firms. In chapter five, Cao examines a novel type of Chinese firm, the born-digitals, by examining their iteration cycle and effects on their speed and learning of internationalization in comparison with the traditional Uppsala model and born-global firms. In chapter six, Mac and Evangelista conduct a survey to investigate the relationship between corporate entrepreneurship and organizational learning as well as their effect on export performance among exporters in Mainland China. It is found out that being entrepreneurial is highly important to Chinese exporters to succeed. But for cultivating such entrepreneurship, it is crucial that the Chinese exporters are able to utilize appropriate an entrepreneurial approach to learning known as international corporate entrepreneurship (ICE) for having a proper knowledge management system to transform learnt knowledge into favorable outcomes.

Part IV, 'Institutions and innovation', contains one conceptual chapter and one empirical study on Chinese integrated resort (IR) operators in Macau. In chapter 7, Redding and Rowley conjecture the challenges faced by China to restructure its economy from a low cost to a high value model at micro and macro-levels. Accordingly, learning and innovation of Chinese firms remain as critical components to deal with greater complexity. In chapter 8, Cuervo focuses on the hospitality and tourism industry, attempting to understand how diversification can be underpinned by innovation. By conducting a multiple case study of three Chinese integrated resort operators in Macau, SJM, Melco, and Galaxy, divergent paths of innovation are identified along with their underlying properties. Finally, we offer our concluding remarks and suggest some directors for future research in part V.

Conclusion

This introduction chapter provides an overview of extant research on the learning and innovation of Chinese firms along the paths of 'Bring In' to 'Go Global'. We have thus witnessed an expanding body of literature with diverse conceptual underpinnings originated from international business, international economics, strategic management and other disciplines. Embarking from this orientation, following chapters included in this edited volume have extended our knowledge about a wide range of mechanisms and processes underlying the learning and innovation of Chinese firms further.

References

Bi, J., Sarpong, D., Botchie, D., & Rao-Nicholson, R. (2017). From imitation to innovation: The discursive processes of knowledge creation in the Chinese space industry. *Technological Forecasting and Social Change*, 120, 261–270.

Björkman, I. & Kock, S. (1995). Social relationships and business networks: The case of Western companies in China. *International Business Review*, 4(4), 519–535.

Blomström, M. & Kokko, A. (1998). Multinational corporations and spillovers, *Journal of Economic Surveys*, 12(2), 1–31.

Buckley, P. J., Clegg, L. J., Cross, A. R., Liu, X., Voss, H., & Zheng, P. (2007). The determinants of Chinese outward foreign direct investment. *Journal of International Business Studies*, 38, 499–518.

Cantwell, J. (2009). Location and the multinational enterprise. *Journal of International Business Studies*, 40(1), 35–41.

Caves, R. E. (1974). Multinational firms, competition and productivity in host-country markets. *Economica*, 41(162), 176–193.

Cheng, C., Hua, Y., & Tan, D. (2019). Spatial dynamics and determinants of sustainable finance: Evidence from venture capital investment in China. *Journal of Cleaner Production*, 232, 1148–1157.

Child, J., & Rodrigues, S. B. (2005). The internationalization of Chinese firms: A case for theoretical extension? *Management and Organization Review*, 1(3), 381–410.

Child, J., & Tse, D. K. (2001). China's transition and its implications for international business. *Journal of International Business Studies*, 32(1), 5–21.

Deng, P. (2009). Why do Chinese firms tend to acquire strategic assets in international expansion? *Journal of World Business*, 44, 74–84.

Dutta, S., Lanvin, B., León, L. R., & Wunsch-Vincent, S. (2021). *Global Innovation Index 2021*. World Intellectual Property Organization, Switzerland.

Dychtwald, Z. (2021). China's new innovation advantage. *Harvard Business Review*, 99(3), 55–60.

EY. (2018). How does geopolitical dynamics affect future China overseas investment? *China Go Abroad* (8th issue). China Mergers & Acquisitions Association.

Findly, R. (1978). Relative backwardness, direct foreign investment, and the transfer of technology: A simple dynamic model. *Quarterly Journal of Economics*, 92, 1–16.

Gaur, A. S., Ma, X., & Ding, Z. (2018). Home country supportiveness/unfavorableness and outward foreign direct investment from China. *Journal of International Business Studies*, 49(3), 324–345.

Grossman, G. & Helpman, E. (1991). *Innovation and Growth in the Global Economy*, Cambridge: M.I.T. Press, 1991.

Hannas, W. C., & Tatlow, D. K. (Eds.). (2020). *China's Quest for Foreign Technology: Beyond Espionage*. London: Routledge.

Javorcik, B. S. (2004). Does foreign direct investment increase the productivity of domestic firms? In search of spillovers through backward linkages. *American Economic Review* 94(3), 605–27.

Li, L. (2004). China's higher education reform 1998–2003: A summary. *Asia Pacific Education Review*, 5(1), 14–22.

Li, S., Easterby-Smith, M., & Hong, J. (2019). Towards an understanding of the nature of dynamic capabilities in high-velocity markets of China. *Journal of Business Research*, 29, 212–226.

Li, J., Strange, R., Ning, L., & Sutherland, D. (2016). Outward foreign direct investment and domestic innovation performance: Evidence from China. *International Business Review*, 25, 1010–1019.

Liao, T., & Yu, C. J. (2012). Knowledge transfer, regulatory support, legitimacy, and financial performance: The case of foreign firms investing in China. *Journal of World Business*, 47(1), 114–122.

Liu, Z. (2001): "Foreign direct investment and technology spillover: evidence from China," *Journal of Comparative Economics*, 30(3), 579–602.

Liu, F. C., Simon, D. F., Sun, Y. T., & Cao, C. (2011). China's innovation policies: Evolution, institutional structure, and trajectory. *Research Policy*, 40(7), 917–931.

Lu, X. (2013). *Dictionary of Management (Chinese version)*. Shanghai: Lexicographical Publishing House.

Luo, Y. & Bu, J. (2018). Contextualizing international strategy by emerging market firms: A composition-based approach. *Journal of World Business*, 53 (3), 337–55.

Luo, Y., Sun, J., & Wang, S. L. (2011). Emerging economy copycats: Capability, environment, and strategy. *Academy of Management Perspectives*, 25(2), 37–56.

Luo, Y., & Tung, R. (2007). International expansion of emerging market enterprises: A springboard perspective. *Journal of International Business Studies*, 38(4), 481–498.

Luo, Y., & Tung, R. (2018). A general theory of springboard MNCs. *Journal of International Business Studies*, 49(2), 129–152.

Luo, Y., Xue, Q., & Han, B. (2010). How emerging market governments promote outward FDI: Experience from China. *Journal of World Business*, 45, 68–79.

Luong, N., Arnold, Z., & Murphy, B. (2021). *Understanding Chinese Government Guidance Funds*. Center for Security and Emerging Technology, March. https://doi.org/10.51593/20200098. Accessed 04.06.2022.

Lyles, M. A., Li, D. & Yan, H. (2014). Chinese outward foreign direct investment performance: The role of learning. *Management and Organization Review*, 10(3), 411–437.

Lyles, M. A., Tsang, E. W. K., Li, S., Hong, J. F. L, Cooke, F. L., & Lu, J. W. (2022). Learning and innovation of Chinese firms along the paths of "Bring In" to "Go Global". *Journal of World Business*, 57(5), 101362.

Ministry of Education of the PRC (2017). Notice from the Ministry of Education and other national governmental departments announcing the list of double first-class universities and disciplines (in Chinese). http://www.moe.gov.cn/srcsite/A22/moe_843/201709/t20170921_314942.html. Accessed 07.06.2022.

MOFCOM (2017). *Regular Press Conference of the Ministry of Commerce*. http://english.mofcom.gov.cn/article/newsrelease/press/201711/20171102667881.shtml. Accessed 12.05.2019.

OECD. (2008). *OECD Reviews of Innovation Policy*: China. OECD, Paris.

OECD. (2020). *FDI in Figures*. http://www.oecd.org/investment/FDI-in-Figures-April-2020.pdf. Accessed 07.05.2021.

PRC State Council (2015). 'Made in China 2025' plan issued. http://english.www.gov.cn/policies/latest_releases/2015/05/19/content_281475110703534.htm. Accessed 07.06.2022.

Rivera-Batiz, L., & Romer, P. (1991). Economic integration and endogenous growth. *Quarterly Journal of Economics*, 106, 531–555.

Rui, H., & Yip, G. S. (2008). Foreign acquisitions by Chinese firms: A strategic intent perspective. *Journal of World Business*, 43(2), 213–226.

UNCTAD. (2018). *World Investment Report 2018: Investment and New Industrial Policies: Key Messages and Overview*. New York and Geneva: United Nations Publications.

UNCTAD. (2021). Global FDI flows down 42% in 2020: Further weakness expected in 2021, risking sustainable recovery. *Investment Trends Monitor*, 38, 1–11. New York and Geneva: United Nations Publications.

Wang, P., Tong, T. W., & Koh, C. P. (2004). An integrated model of knowledge transfer from MNC parent to China subsidiary. *Journal of World Business*, 39(2), 168–182.

Wei, Y. & Liu, X. (2006). Productivity spillovers from R&D, exports and FDI in China's manufacturing sector, *Journal of International Business Studies*, 37, 544–557.

Weinstein, E. (2022). Beijing's 're-innovation' strategy is key element of U.S.-China competition. Brookings. https://www.brookings.edu/techstream/beijings-re-innovation-strategy-is-key-element-of-u-s-china-competition/. Accessed 30.05.2022.

Williamson, P. J., & Yin, E. (2014). Accelerated innovation: The new challenge from China. *MIT Sloan Management Review*, 55(4), 27–34.

Yip, G. S., & McKern, B. (2016). *China's Next Strategic Advantage: From Imitation to Innovation*. Cambridge, MA: MIT Press.

Zhou, C., Delios, A., & Yang, J. Y. U. (2002). Locational determinants of Japanese foreign direct investment in China. *Asia Pacific Journal of Management*, 19(1), 63–86.

Zhou, D., Li, S., & Tse, D. K. (2002). The impact of FDI on the productivity of domestic firms: the case of China. *International Business Review*, 11(4), 465–485.

Part II: **Knowledge Transfer and Managerial Cognition**

Yixin Sun

2 Reverse Knowledge Transfer from Developing Countries to Chinese Multinational Corporations: The Case of Haier

The market in developing countries is the source of innovation.
– Managing Director (Haier India)

Introduction

A greater amount of research attention has been devoted towards the emerging market multinational corporations (EMNCs) that integrate resources internationally in comparison to the developed MNCs who hold a headquarter-centric perspective of knowledge transfer (Liu & Meyer, 2020). EMNCs are said to hold similar features as that of MNCs when it comes to acquiring technological knowledge and managerial expertise from subsidiaries (Child & Rodriguez, 2005), which eventually helps them gain a competitive position among global companies (Luo & Tung, 2007). EMNCs, especially Chinese MNCs, require their overseas subsidiaries to become independent units and survive using their own capabilities in order to provide them a high level of autonomy to carry out different value activities. Such autonomy allows the subsidiaries to tap into local knowledge (Andersson et al., 2002), which thus, enhances their product innovation and knowledge development (Persaud, 2005; Phene & Almeida, 2008). Consequently, instead of being recipients, subsidiaries of EMNCs have transformed their roles to become knowledge creators and have come to reverse transfer knowledge back to their headquarters (HQs), improving their HQs' overall performance (Liu & Meyer, 2020).

However, subsidiaries playing the role of agents may not voluntarily engage in reverse knowledge transfer (RKT) in accordance to their HQs' interests, such as integration of knowledge from the principal-agency perspective of the EMNCs (Konga et al., 2018). As excludable knowledge is crucial for having success in globally competitive markets, all independent subsidiaries protect their competitive advantage and prevent themselves from being involved in the process of RKT (Gupta & Govindarajan, 2000). Although RKT characterized by intensive information exchange is highly beneficial to HQs, it may prove to be a time-consuming process for the subsidiaries holding complex knowledge and the interpretation of specific contexts (Andersson et al., 2002). As more and more autonomous subsidiaries are speculated of having lesser hierarchical cooperation with their HQs (Noorderhaven & Harzing,

https://doi.org/10.1515/9783110715002-002

2009), RKT has emerged as a problematic issue in the case of EMNCs' acquisitions, during when, HQs may encounter extreme resistance from acquired subsidiaries for making efforts toward RKT (Ciabuschi et al., 2017). Hence, an independent subsidiary may be less willing to participate in RKT, which is in line with the previous research finding that subsidiary autonomy has a negative impact on reverse knowledge flows (Gammelgaard et al., 2004; Lord & Ranft, 2000).

With there being such an interesting paradox between HQs' objective to facilitate knowledge integration and subsidiaries' desire for a higher level of autonomy, an exclusive tool for knowledge integration called coordination mechanisms, including personal and electronic means, has been extensively adopted by EMNCs as a solution. By studying the case of Haier, we aim to explore how EMNCs can make use of personal coordination mechanisms (PCM) (e.g., face-to-face meetings, expatriates transfer) and electronic coordination mechanisms (ECM) (e.g., emails, information system) to mitigate the negative impact of subsidiary autonomy on RKT. Established in 1984, Haier is a world-leading multinational company with continuous and growing innovation in the global market. With the support of coordination mechanisms, Haier has gone through a pathway of internationalization by entering traditionally mature markets such as Europe, the USA, and Japan, as well as emerging markets such as India, Malaysia, and Pakistan. Thus, the home country and industry contexts of Haier provide a nuanced understanding of the complex relationships between subsidiary autonomy, coordination mechanisms, and RKT.

Using case studies with the purpose of theory development (Eisenhardt & Graebner, 2007), we make two contributions to the literature on EMNCs. First, this study extends the prior work on subsidiary autonomy and RKT, and brings out how a high degree of subsidiary autonomy acts as a hindrance to RKT. This study further highlights the alignment of subsidiary autonomy and coordination mechanisms to articulate the underlying RKT processes of the different value activities that operate between HQs and their subsidiaries, indicating that the coordination mechanisms moderate the relationship between high subsidiary autonomy and RKT. In the case of Haier, a negative association is mitigated when the coordination mechanisms are utilized. Second, by using a typical inductive research method, this study sheds light on the sketch of an emerging RKT paradigm in the context of EMNCs venturing into other developing countries, wherein the PCM facilitates the transfer of subsidiaries' technology as well as their integration with HQs, and the ECM ensures efficient communication and enhances management of overseas subsidiaries.

The study is organized as follows. The second section provides a brief literature review of RKT, subsidiary autonomy, and coordination mechanisms. In the following sections, we describe the company background of Haier and the three case vignettes as the primary findings. Finally, we discuss future research directions and conclusions.

Literature Review

Reverse Knowledge Transfer

Research on knowledge transfer has enjoyed a long tradition. Extensive literature on knowledge transfer within MNCs has identified knowledge flow as unidirectional, moving from the HQ in the home country to its overseas subsidiary. However, in recent years, an increasing number of studies have analyzed RKT – the transfer of know-how from an overseas subsidiary to their headquarters (Ambos et al., 2006; Frost & Zhou, 2005; Yang et al., 2008). In Gupta and Govindarajan's (2000) pioneering research, RKT in a cross-border scenario has been conceptualized as the knowledge and skills that flow from overseas subsidiaries to their headquarters in the home country. Minbaeva et al. (2003) have refined the definitions and empirical evidence for RKT, with Ambos et al. (2006) providing the critical caveat that the vital element in reverse knowledge transfer is "not the underlying (original) knowledge, but rather the extent to which the receiver acquires potentially useful knowledge and utilizes this knowledge in their operations" (Minbaeva et al., 2003, p. 587). The term "reverse" is used to carry out the distinction between the conventional form of "forward" transfers from headquarters to subsidiaries (Yang et al., 2008) and the "lateral" transfers between subsidiaries (Ambos et al., 2006). A suggestive definition of RKT here is exchanging the know-how from the international subsidiaries in host countries to the HQs in home countries (Noorderhaven & Harzing, 2009).

Scholars have widely acknowledged that the utilization of reverse knowledge transfer may benefit HQs (Eden, 2009; Michailova & Mustaffa, 2012), which contributes critically to creating firms that have competitive advantages (Ambos et al., 2006). Under such circumstances, reversely transferred knowledge from subsidiaries is characterized by the seeds of enhanced knowledge found within HQs (Bartlett & Ghoshal, 1989), thus, leading to the sources of innovation. Aside from the research on the beneficial effects of RKT, scholars have utilized a wide variety of theoretical perspectives to find the various determinants of RKT (e.g., Yang et al., 2008). For example, based on the knowledge relevance theory, Yang et al.'s (2008) theoretical ideas on RKT show that HQ's capability to adopt knowledge overlap is positively associated with RKT. Consistent with the factors in communication theory, Gupta and Govindarajan (2000) have conceptualized the knowledge flowing out of a subsidiary in their overarching theoretical framework, offering a starting point for the consideration of the existence and richness of the transmission channel as one of the main impact factors of RKT.

The emphasis on RKT within knowledge-based views has declined as the assumption underlying the knowledge-based theory demonstrated knowledge to already exist in the boundary of each subsidiary unit (Najafi-Tavani et al., 2012). Nevertheless, some scholars have argued that a subsidiary's knowledge base does not always exist, and that it is determined by the level of autonomy the subsidiary has, serving as a critical determinant of the knowledge created via intensive interaction with local markets (Foss & Pedersen, 2002; Cantwell & Mudambi, 2005). In other words, subsidiary autonomy may affect knowledge creation as an origin of the entire RKT process. When a greater degree of autonomy is granted to the subsidiary, it is believed that the relationship between HQ and the subsidiary will be less hierarchically connected and more independent (Noorderhaven & Harzing, 2009), thereby increasing the dilution of reciprocal trust (Rabbiosi, 2011) and MNE's autonomy-control tension (Asakawa, 2001), which will be negatively related to the reverse transfer of knowledge from subsidiaries and might not allow MNCs to benefit from the knowledge held by subsidiaries (Minbaeva, 2007).

Consequently, the current literature has come to capture the coordination mechanism ties in RKT (Gupta & Govindarajan, 2000; Noorderhaven & Harzing, 2009; Schulz, 2001; Tsai, 2001). For example, shared vision and goal convergence among the top managers in subsidiaries and HQs can be stimulated via coordination mechanisms (Tsai & Ghoshal, 1998), which serve as a means of applying more social control to sustain mutual trust and modify the autonomy-control tension found in the HQ–subsidiary relationship (Nobel & Birkinshaw, 1998). Eventually, this manifests as a growing pattern of RKT. For instance, while Gupta and Govindarajan (2000) found a positive relationship between coordination mechanisms and knowledge outflows from the identical subsidiaries to HQs, they failed to consider the different levels of subsidiary autonomy jointly. The pioneering scholarly research on the combined effect of subsidiary autonomy and coordination mechanism has its roots in Noorderhaven and Harzing's (2009) classic work that links the negative impact of autonomy with the application of social interaction, showing how social interaction leads to the creation of considerable knowledge in reverse knowledge flows for highly autonomous subsidiaries. In the context of this article, we analytically conclude that the role of the coordination mechanism, as emphasized by Gupta and Govindarajan (2000) as a central determinant of RKT, has mitigated the negative effect the presence of highly autonomous subsidiaries has on RKT.

Subsidiary Autonomy

Subsidiary autonomy refers to "the constrained freedom or independence available to or gained by a subsidiary, which enables it to make certain decisions on its own behalf" (Young & Tavares, 2004). In comparison to O'Donnell (2000), who categorized subsidiary autonomy into strategic and operational decision-making authority, Chiao and Ying (2013) came up with a categorization whose scale and scope have a more comprehensive dimension. The scale represents the magnitude of the decision-making rights that subsidiaries own for value activities (Chiao & Ying, 2013). The term "scope" refers to the number of value activities (Porter, 1985), namely production, marketing, personnel, finance, and R&D, which the subsidiaries can decide upon. These are in line with the conceptualized dimensions showed by studies, such as Bowman et al. (2000) and Edwards et al. (2002). More recently, Cavanagh et al. (2017) established a more concise dimension that was categorized into administrative functions autonomy, production autonomy, innovation autonomy, and sales autonomy. In the context of this article, subsidiary autonomy refers to the limited power or independence with two dimensions, namely scale and scope, possessed by subsidiaries, allowing them to make decisions pertaining to different value activities efficiently and flexibly (Chiao & Ying, 2013).

Beginning with an early focus on subsidiary autonomy, a body of research has come to offer indicative information to map autonomy as an outcome in the subsidiary (Taggart, 1996; Young & Tavares, 2004). For example, Young and Tavares (2004) developed a conceptual approach that has come to have a profound influence on the understanding of well-supported affecting factors of subsidiary autonomy. According to Young and Tavares (2004), subsidiary autonomy has five determinants; authority, expatriate managers' motivation, power with head office, and trust between HQs and subsidiaries. Researchers working in the respective area have come to realize that subsidiary autonomy extends beyond the existence of an outcome in the HQ-subsidiary relationship, and have thus, started conceptualizing subsidiary autonomy as an input that is dependent on the other factors in the HQ-subsidiary relationship, including knowledge flows (Gupta & Govindarajan, 2000), human resource management (HRM) operation (Fenton-O'Creevy et al., 2008), subsidiary performance (Gammelgaard et al., 2012), and RKT in MNCs (Rabbiosi, 2011).

There has been a debate on the relationship between the different levels of subsidiary autonomy and the extent of RKT. Adopting an agency theory perspective, previous empirical studies have discovered subsidiaries with autonomy to resist HQs' control, with this resulting in tensions between them due to lesser mutual trust (Saliola & Zanfei, 2009). Therefore, it is not surprising that autonomous subsidiaries have no intention of engaging in the RKT process, which may create value for the HQs. In other words, a higher degree of subsidiary autonomy negatively influences the transfer of knowledge from the subsidiaries to HQs (Gammelgaard et al., 2004). However, this is at odds with Tsai's (2002) finding of positive impact. The consequence of this oversight is the inadequate recognition of coordination mechanisms utilized within the HQ-subsidiary dyads (Rabbiosi, 2011). As Asakawa (2001) suggests, if subsidiaries have too much autonomy in their local markets, they may have some possibilities to become distracted from their HQ's objectives, which in turn can result in conflict and distrust. If not managed well, the target incongruence between the HQs and subsidiaries caused by a high level of autonomy will result in less motivation for RKT, which does not seem to facilitate the process of RKT (Najafi-Tavani et al., 2015). More importantly, such an examination has mainly remained at the MNC level. Research then needs to progress in order to investigate the EMNC level by shifting from one channeling process to more value activities in conceptualized dimensions (Monteiro & Birkinshaw, 2017). This would provide interesting viewpoints that the level of subsidiary autonomy may vary across value activities (Collinson & Wang, 2012; Rugman et al., 2011), which would in turn help EMNCs identify the different levels of autonomy found among value activities and how they influence RTK to make use of the knowledge transferred to HQs.

Coordination Mechanism

Coordination mechanism for long have been a topic of interest. The evolution of coordination mechanisms has moved from there being an administrative tool used for integrating the different units within multinational corporations (Martinez & Jarillo, 1989) to there being an exclusive tool used for enhancing knowledge transfer between the units within MNCs (Rabbiosi, 2011). Typically, studies have concluded the existence of two distinct categories, including PCM and ECM, to explore this issue (Ambos & Ambos, 2009; Rabbiosi, 2011). First and foremost, PCM comprises of different personal interactions such as face to face meetings, expatriate transfers, and committee activities (Wang et al., 2019; Zhou et al., 2020). These interactions build a "verbal network" that transmits complex knowledge and complements tacit information in the written systems (Edstrom & Galbraith, 1977). In addition, these interactions might moderate cognitive hindrance and create trustworthy relationships (Rabbiosi, 2011). Second, ECM comprises of different technical media such as emails, phone conferences, and more sophisticated electronic systems (Jean & Sinkovics, 2010; Luo & Bu, 2016). These media function as an effective platform for RKT. Following the recent work by Kane and Alavi (2007), ECM helps reduce knowledge heterogeneity via the codification and integration based on exploitation learning.

It has recently been acknowledged that coordination mechanism has been a vital dimension of RKT to effectively overcome the significant impediments and efficiently enhance the process of knowledge flows found between the subsidiaries and HQs (e.g., Wang et al., 2019). In this regard, coordination mechanisms play a two-fold role. The first is to provide efficiency in knowledge management to overcome the inherent disadvantages accompanying a high level of subsidiary autonomy (Rabbiosi, 2011). From a cognitive perspective, PCM helps build shared values and trust among the managers of HQs and subsidiaries (Rabbiosi, 2011). Following this, HQs will attempt to access and leverage locally embedded knowledge from every subsidiary through coordinated approaches that effectively help them cope with the various knowledge types and diversified value activities undertaken by their subsidiaries. The second is to eliminate the impediments to the different subsidiary knowledge types. For example, PCM enriches the information transmission channels where tacit knowledge from subsidiaries can be transferred through a cognitive platform (Gupta & Govindarajan, 1991). In addition, the ECM is beneficial for less codified knowledge from subsidiaries, including standard information and well-structured data. With the support of "knowledge objects" through which reverse knowledge can be collected and used, ECM contribute toward the achievement of the "people-to-documents" process (Wang et al., 2019), i.e., employees from MNE units may easily hold and manipulate essential information (Rabbiosi, 2011). To conclude, the right selection of coordination mechanism is one of the core dimensions of company strategy, and more importantly, it has been examined to be vitally relevant to explain RKT phenomena within MNCs (Schulz, 2001; Tsai, 2002).

As internationalization continues to develop, increasing volume of research have come to lay emphasis on EMNCs, which serve as latecomers in the globally competitive market. Although the importance of the coordination mechanism for RKT has been accepted under the topic of MNCs, it appears to be vacant in EMNCs that have differences in comparison to the MNCs. The typical features of EMNCs, such as weaker proprietary advantages and home institutions (Bartlett & Ghoshal, 1989), have determined their particular strategy, namely obtaining critical knowledge and capabilities in the host country and setting up their competitive positions at the world stage (Luo & Tung, 2007; Deng, 2009; Kumar et al., 2019; Luo and Bu, 2018), to be in line with Buckley et al.'s (2007) finding that Chinese MNCs often choose to enter foreign markets with rich technological endowments. However, prior research has predominantly highlighted the process of RKT adhered by subsidiaries in developed economies, focusing on the little, if any, differences in the knowledge of subsidiaries in developing economies. For instance, Liu and Meyer (2020) have pointed out that the condition of RKT in Chinese MNCs may be different for the subsidiaries operating in developed countries and those operating in developing countries.

It is worth noticing that the delegation of significant autonomy to overseas subsidiaries will negatively affect RTK. Nevertheless, previous research has apparent limitations, in that, they have failed to consider the interplay between subsidiary autonomy and coordination mechanisms concerning RKT. Accordingly, recent research studies have shown EMNCs, particularly Chinese MNCs, to adopt a "light-touch approach" while managing their overseas subsidiaries, and thus, providing them a high degree of autonomy (Wang et al., 2014; Chen et al., 2018). Subsidiaries in developing economics reap the benefits of their autonomy to interact extensively with local markets for knowledge creation and product innovation. Meanwhile, HQs leverage this reverse knowledge depending on the proper coordination mechanism underlying this strategy. However, research to date have revealed little about the underlying RKT processes through which highly autonomous subsidiaries achieve product innovation and knowledge integration with HQs. Consequently, it is vitally important to pay more attention to the successful cases of RKT in cross-border subsidiaries of Chinese MNCs (Liu & Meyer, 2020) and further investigate how they facilitate different coordination mechanisms to increase the accumulation of overseas knowledge and innovation. In our context, an explicitly absent recognition is the combined effect of the EMNE-level autonomy and coordination mechanisms that affect the RKT of subsidiaries in Chinese MNCs: *How do the autonomous subsidiaries of Chinese MNCs in developing countries reverse transfer knowledge in conjunction with the different coordination mechanisms to their HQs?*

Methodology

In this study, we have carried out multiple case studies via the use of theoretical sampling that employs the logic of replication (Yin, 1984). We have illustrated three different case vignettes about the Haier subsidiaries in developing countries following the year of inception presented in Table 2.1. In the cases of all the three subsidiaries located in developing countries, RKT prominently features the stated objectives, which helps to "maximize opportunities to discover variations among concepts and to densify categories in terms of their properties and dimensions" (Strauss & Corbin, 1998, p. 201).

Table 2.1: An overview of Haier's three foreign subsidiaries.

Case	Year of establishment	Entry mode	Location	Role
Subsidiary A	2000	Joint venture	Pakistan	Production center and laboratory
Subsidiary B	2004	Greenfield investment	India	Production and R&D center
Subsidiary C	2012	acquisition	Malaysia	Production center

In this study, we also used a stepwise process with different sources to identify the three overseas subsidiaries of Haier located in developing economies. As the first step, we gathered relevant information by searching news, press releases, and company announcements on Haier's official website (www.haier.com). As the second step, we leveraged the researcher's professional and personal networks, thereby seeking out interviewees in the identified vignettes and conducting semi-structured in-depth interviews with senior managers. This helped us to collect information about the mitigating effect of coordination mechanisms on RKT in highly autonomous subsidiaries. Table 2.2 provides a description of the primary interviewees from the three subsidiaries and the HQ of the Haier Group. Accordingly, we stopped increasing the number of interview informants when further communications had ceased to generate additional significant information. Data collection using a theoretical sampling approach enabled the overall vignettes to hold complementary perspectives toward the RKT from Haier subsidiaries to their HQ, thus enhancing the construct validity of data collection.

The interviews were characterized by open-ended questions asked through WeChat's voice calls. The aim was to ask interviewees about their working experience

Table 2.2: Interviewees.

Case	Number of interviews	Positions
Subsidiary A	2	Senior project manager
		Indian marketing manager
Subsidiary B	2	Pakistan managing director
		Senior project manager
Subsidiary C	3	Malaysia managing director
		Malaysia marketing managers
HQ	3	Overseas HR managers

covering three keywords, namely subsidiary autonomy, coordination mechanism, and RKT, to take advantage of the fact that most of the interviewees from subsidiaries had been involved in the daily knowledge management and the RKT process with the HQ.

Case Background

History of Haier

During the stage of internationalization (1998–2005), Haier had its sights on the worldwide market by having "RenDanHeYi" as its inspiration. Through this, Haier went on to build an overseas network of distributors and an after-sales service system. A major milestone was the Haier American Factory, with the strictest entry standards found in the year 1999. Following the first successful exploration, Haier simultaneously shifted its attention to the less developed markets that were potentially underestimated. Established in 2000, Haier Pakistan is a joint venture where the Haier Group has a share of 55%, and the Ruba Group has a share of 45% to date. As a solid proof of Pak-China friendly relationship, the Pakistan Haier-Ruba Economic Zone was the first Chinese overseas trade and economic cooperation, which came into being with the approval of the Chinese Ministry of Commerce and the Pakistani government (Haier official website, 2021b). Starting in 2004, Haier established a wholly owned subsidiary in India. Stemming from Haier's philosophy of "Inspired Living," Haier India was best known for best-in-class innovations comprising of bottom mounted refrigerators (BMR) with convertible sections, which took into consideration the demands of the Indian customers. Shortly afterwards, Haier Malaysia was set up. Starting its operations in 2005, Haier Malaysia came to offer innovative products to meet the demands of the customers. Intending to become a leader of the white appliance business in Southeast Asia, Haier acquired Sanyo Electric's consumer electric appliances business comprising of refrigerators and washing machines in 2012. This laid a solid foundation for Haier's ever-growing sales growth (Haier official website, 2021a).

Due to cultural distance and diversity, the Haier Group only plays a central role in tracking and monitoring the final performance of each subsidiary. A similar management strategy of surviving on subsidiaries' abilities is applied to each subsidiary in Haier, regardless of the distinction of their host country. Various types of autonomy, namely production, marketing, personnel, finance, and R&D, are granted to most of the Haier's subsidiaries in developing countries. As a result, while these subsidiaries utilize their autonomy and interact extensively with local markets to upgrade the products and meet the customers' demands, the HQ of Haier leverages this reverse knowledge and information depending on the proper coordination mechanisms underlying this strategy.

Haier's Reverse Knowledge Transfer from Foreign Subsidiaries

Subsidiary A. Haier Pakistan is a joint venture. On 16th November, 2019, Haier Pakistan initially opened an experiential-service store called "Smart Home" in downtown Lahore. With the built-in "Smart Home" store, customers not only experienced smart appliances firsthand with the assistance of highly trained staff, but were also asked to share their opinions about this human-machine scenario, which were recorded by the customer service teams and translated into documents by the specific technical groups in Haier Pakistan. Shortly afterward, the technical staff might have conducted physical experiments in their local laboratory to develop "small features" innovation by taking into consideration the needs of Pakistanis. This type of innovation included structural laboratory tests based on the customers' responses to "Smart Home," which is transferred via Haier internal emails for further high-tech integration in the HQ's R&D center. It then becomes a part of the HQ's knowledge inventory. The list of productive knowledge on relevant technical staff will be filled in the system applications and products in data processing (SAP) and warehouse management system (WMS) for information transformation. At the same time, periodic meetings with the expatriates nominated by the Haier Group are held three times a week, wherein the expatriates report their latest progress in product innovation and business campaigns. Every year, the Haier Group holds an international meeting with the directors and some of the key technical staff from Haier Pakistan on the 20th of September. This meeting functions as a platform for every subsidiary to share their annual performance and localized innovation with the HQ.

Haier Pakistan has a high degree of subsidiary autonomy with regard to manufacturing, selling, and HRM in order to survive on its own abilities. The HQ of the Haier Group turns its attention to this single subsidiary only to check if the annual turnover of the subsidiary has spectacularly reached the stated business volume since the beginning of the year. Following this logic, the primary task of Haier Pakistan is to make the most of discovering the potential local markets and obtaining more market shares, rather than depleting the many human resources to transfer a great deal of knowledge. Alternatively, due to cultural diversity, autonomous local employees face difficulties comprehending the HQ's targets of RKT, resulting in them focusing more on achieving their targets. Even if they have related their creative thoughts, they do not know if the pieces of knowledge are helpful to HQ and how to interpret them into standardized information correctly, as the product standards and specifications are almost distinctive between China and Pakistan.

Taken as a whole, the expatriates burdened with consistent goals from the HQ are sent by the Haier Group to join the senior management team of Haier Pakistan, who serve as a bridge of shared vision. In brief, they not only have been given the responsibility to make autonomous decisions with regard to retailing and basic HRM, such as employee recruitment and salary formulation, but also have been given key missions to maintain the HQ's spirit and help overcome the goal incongruence arising from cultural barriers. Moreover, PCMs such as face-to-face meetings play a vital role in both technological and cultural integration, which encourages mutual trust and harmonious relationships between Pakistani employees and HQ employees. In contrast, the use of ECM mitigates the knowledge heterogeneity arising due to regional differences in the product standards and customer demands based on extensive localized interactions. The staff in HQ may take advantage of knowledge codification and exploit it through the ECM. To be precise, computer application software such as SAP and WMS offer an efficient channel to exchange knowledge without consuming a significant energy of the subsidiary.

Subsidiary B. Haier India is a wholly owned subsidiary of the Haier Group, with its own manufacturing unit and R&D center. This subsidiary has significant autonomy with regard to designing and upgrading the production lines. To adapt to the necessities of contemporary Indian families, the R&D center of Haier India has conducted extensive marketing research to discover the Indian customers' appetites for better home appliances. For example, most Indian women are required to bend down to access the freezer at the bottom of traditional refrigerators. To ensure that the Indian customers bend only to access the lesser commonly used things in a fridge, Haier India came up with a bottom-mounted refrigerator (BMR) technology, wherein the refrigerator is designed in such a way that the more used freezer is at the top and the less-used refrigerator is at the bottom, leading to a 90% decrease in customers' bending (Kanchwala, 2021). In terms of ECM, Haier India has transferred its innovative structural drawing of BMR technology combined with visualized Indian technology requirements through internal emails to the HQ's technical reserve. Meanwhile, the financial supply chain adjusting to BMR technology will be added to the SAP and Global Entity Management System (GEMS), which solves how the HQ and its subsidiaries manage the costs of raw material, overseas tax, and financial support. In comparison to Haier

Pakistan, the top executives in Haier India also comprise of qualified expatriates from the HQ who report about the innovative technologies adopted by the subsidiary in the regular meetings held in the HQ. Due to the Covid-19 pandemic, all face-to-face meetings were replaced by periodical written reports and online meetings via Zoom.

Undoubtedly, autonomous Haier India (with its steadfast capabilities) tends to protect BMR technology as one of its competitive advantages to maintain an industry-leading position as the top executives in India are always adept at leaving an appropriate margin for unforeseen circumstances. As such, aided by the autonomy granted by Haier Group, Haier India might have been less interested in sharing excludable knowledge, which has prevented them from engaging in the RKT process. Furthermore, if Haier India is involved in RKT, they are required to make heroic efforts to manage exceptional knowledge concerning the existence of a different protective system of intellectual property rights between India and China. Similar to the situation in Haier Pakistan, owing to the cultural impediments and product specifications, Haier India needs considerable time and energy to interpret and codify the specific context and complex knowledge based on Indian markets, limiting its involvement in RKT.

The technological knowledge, namely BMR, in Haier India has a dominant characteristic of stickiness, which should be enriched through a more cognitive platform; thus, PCM such as online and offline meetings among Indian R&D staff and HQ staff are beneficial for transferring codified knowledge. If any party has met with misunderstandings, they can use this verbal system to learn to deal with how to store customers' personalized food items and smartly cater to their usage patterns. In addition, the expatriates, who work as managing directors in Haier India, are motivated by the incentives provided by HQ, whose spirits and missions are congruent with the HQ's target. They are responsible for practicing Haier's business model in the Indian subsidiary where the Indian employees might have a sense of belongingness to Haier culture and might be trying their best to fulfill their jobs' missions. Furthermore, the expatriates assist in bringing back the most valuable part of the BMR to the HQ by determining the intellectual property laws and arranging the duration of technological protection. As a result, the Haier Group can consider designing perfect convertible refrigerator modes for the younger Chinese generation. With regards to ECM, Haier India uses computer software like SAP and GEMS, which functions as a "screen" in which different product specifications can be easily unified. As a result, the HQ's managers can monitor the daily operations of the Indian subsidiary, which yields complete self-learning about all the latest implemental progresses made by the subsidiary. More importantly, efficient software provides an opportunity to collect information with less time and energy; likewise, the HQ's managers access and manipulate practical knowledge through the "people-to-documents" mechanism, eliminating the impediments of space and distance.

Subsidiary C. After acquiring the Japanese brand Sanyo in Southeast Asia in 2012, the Haier Group is a more completed business organization. It has two R&D centers, four manufacturing units, and six localized marketing centers in Southeast Asia (Haier official website, 2021a). Haier Malaysia is one of the key localized marketing centers. It has an entire customer care system and after-sales service. The RKT in this case is divided into two types. One is through PCM, particularly for Sanyo Electric Corporation's leading technology relating to energy usage, which has been deployed and upgraded to Haier Malaysia's product innovative offerings along with the management models used for integrating Haier's business philosophies and Japanese culture inherited from Sanyo Electric Corporation. In addition, the expatriates and key technical staff involved in Sanyo product design return to the HQ at least three times a year for technological and managerial knowledge exchange. In addition, the periodic reports on acquisition continuity and dual-brand strategy, including "Haier" and "Aqua" in the entire operation across Malaysia, will be frequently discussed via online meetings and internal emails to the HQ. The other is through ECM, whereby the experimental data linking with energy storage and warehouse systems are transferred through internal emails and computer software such as SAP and WMS.

Culturally diversified situations can be captured in the same way in Haier Malaysia, in which local employees under highly autonomous management appear to be easily indolent toward their work, let alone implementing a time-consuming process of interpreting excludable and complex knowledge for the HQ. In addition, influenced by the subconscious cultural orientation, Japanese employees with considerable autonomy in pre-acquisitions concentrate on the subsidiaries' development and thus, do not have consistent targets along with the HQ. The problems of intellectual property rights, in a similar manner, should also be considered in Haier Malaysia's RKT process, with there being a distinction in protective laws between China and Southeast Asia. It is similar in the case of Haier India, which wants to protect its steadfast advantages rather than sharing its knowledge with the HQ and being involved in RKT activities, posing a threat of losing domain power in the entire market.

Taken together, the Haier Group makes most of the ECM and PCM moderate the negative impact of subsidiary autonomy. The ECM are far beyond the effective transmission channel, regardless of the distance between the host country and the home country. They have come to enable a hierarchically new cooperation between the Haier Group and Haier Malaysia. As a monitoring screen at the HQ, SAP plays an essential role of managing all the employees engaged in the previous Sanyo Electric's businesses. When Haier Malaysia in the post-acquisition stage drafted new principles of HRM focused on its local region, the recruitment information was recorded and transferred efficiently through SAP to HQ's HR center, leading to personnel resource construction and reconstruction. All the raw material purchases related to new product innovation are listed on SAP; likewise, warehouse information about the local markets will be tracked via WMS. Alternatively, PCM reduces the hindrance of cultural differences through oral transmission, thereby contributing to target congruence and trustworthy relationships among the staff of both the parties. Furthermore, it mitigates the knowledge heterogeneity of formal technological meetings held by the Haier Group. For example, technical staff introduce energy-efficient appliances and exhibit how the vibrating operations remove stains without damaging the clothes, which is perceived as an excellent opportunity to enhance the comprehension of the HQ's staff. As the managing directors in Haier Malaysia, expatriates can manage local employees through the HQ's spirit and goals. They can also formulate employees' salaries and long-term incentive systems in accordance to the local markets. By integrating Haier's business philosophies and the management culture inherited from Sanyo, it will be possible to have employees holding objectives that are consistent with that of the HQ's.

Even though Haier's subsidiaries from less developed economies have considerable autonomy, they have interactively engaged in the RKT process by implementing proper coordination mechanisms. In Table 2.3, we have presented the links between the characteristics of all the critical factors in the three case vignettes. While the second column of Table 2.3 shows the identified factors that are negatively influenced by

Table 2.3: Comparison of three Haier subsidiaries in developing countries.

Case	Subsidiary autonomy	Coordination mechanism	Mitigating effects on RKT
Subsidiary A	Draw attention toward the annual turnover, target the incongruence with the HQ, have no ideas of which knowledge is practical and how to interpret these knowledges	PCM: expatriates, formal face-to-face meetings, online meetings in times of Covid-19, periodical reports ECM: Internal emails, SAP, WMS	PCM: expatriates are sent by the Haier Group with the consistent goals of delivering the spirit of the HQ, overcoming goal incongruence arising from cultural barriers, and ensuring the mutual trust and harmonious relationship between the Pakistani employees and the HQ's employees ECM: mitigate knowledge heterogeneity pertaining to product standards, take advantage of knowledge codification and exploitation, open an efficient channel without consuming much energy.

Table 2.3 (continued)

Case	Subsidiary autonomy	Coordination mechanism	Mitigating effects on RKT
Subsidiary B	Protection of competitive advantages, leaving an appropriate margin for unforeseen circumstances, problems of intellectual property right, a lot of time and energy is needed to interpret and codify	PCM: expatriates, formal face-to-face meetings,online meetings in times of Covid-19, periodical reports ECM: internal emails, SAP, GEMS	PCM: a verbal system to ensure understanding, expatriates' responsibilities to practice Haier business model in the Indian subsidiary, a sense of belongingness toward Haier culture and fulfilling the missions of their jobs eventually, figuring out the intellectual property laws and arranging the duration of technological protection ECM: a particular "screen" to complete further self-learning about the latest implementation progress, collecting reverse information with less time and energy, eliminating impediments of space and distance
Subsidiary C	Easy indolence, subconscious cultural orientation, problems of intellectual property right, protection of steadfast advantages	PCM: expatriates, formal face-to-face meetings,online meetings in times of Covid-19, periodical reports ECM: internal emails, SAP, WMS	PCM: reduce the hindrance of culture differences, target congruence and trustworthy relationship among both the parties' staff, training and management of local employees in alignment with the HQ's spirit and goals ECM: setting up of a hierarchically new cooperation, setting up of a monitoring screen to maintain the performances

subsidiary autonomy, the last column of Table 2.3 illustrates how the different coordination mechanisms moderate RKT between the Haier Group and its subsidiaries.

Conclusion

To conclude, this study has presented three inspiring cases of cross-border subsidiaries to bring out the combined effect of EMNE-level autonomy and coordination mechanisms on the RKT process in Chinese MNCs. In brief, this study makes several theoretical contributions to the literature. First, because of some neglected insights pertaining to RKT in previous research (for reviews, see Chung, 2014; Liu and Meyer, 2020), this study explores how the Haier Group properly leverages an intervening factor, namely coordination mechanisms, to manage the autonomous operations of overseas subsidiaries in the light of different target regions and culture using an empirically accurate view. Furthermore, this study supports the potential premise of the negative impact of high autonomy that subsidiaries may be distracted from their HQ's prime targets, leading to less motivation for RKT and distrust relationships between the HQ and subsidiaries (Rabbiosi, 2011). In addition, the abovementioned interviews also show how the other specific contingencies such as indolent styles and leaving adequate leeway for their competitive knowledge are caused by a high degree of subsidiary autonomy. Along this line of reasoning, this study stresses on the need to have more awareness about the conjunction of subsidiary autonomy and the buffering effect of proper coordination mechanisms by explaining how a successful Chinese MNC has solved this problem. Overall, the findings of this study expand our knowledge of RKT within the different coordination mechanisms and offer a possible solution for the HQs to solve the conundrum of managing autonomous subsidiaries.

Second, this study has attempted to substantiate and contribute to RKT literature by formally examining how EMNC-level reverse knowledge, rather than MNC-level reverse knowledge, can shift from the developed countries to less developed countries (Meyer and Liu, 2020). This shift has resulted in the knowledge accumulation of product innovation and managerial practices in EMNCs based on secondary data and the interviews held with Haier employees. While previous literature has mainly addressed how the coordination mechanisms help reduce the heterogeneity of subsidiary knowledge as well as increase the efficiency of information transmission, this study extends the merits of coordination mechanisms. Specifically, PCMs such as expatriates provide an entirely verbal system and thereby, decrease the hindrance of goal incongruence caused by cultural differences, leading to there being a consistent spirit and harmonious relationship between the subsidiaries and HQs. In contrast, the ECM, including software, ensure that there is a hierarchically new structure to monitor and maintain the RKT process, which eliminates the impediments of consuming a lot of time and energy. Although each subsidiary of Haier has its peculiarity, it exhibits similarities and applies collaborative practices with the HQ, evolving into a mature RKT pattern.

Since this study drew upon the cognitive perspectives of coordination mechanisms to investigate how it facilitates the RKT process in highly autonomous subsidiaries, it has limitations that may inform further research. As found by the other studies on RKT (Chung, 2014; Rabbiosi and Santangelo, 2013), a high degree of subsidiary autonomy does not always equate to the failure of RKT. Although this study proposes that coordination mechanisms as a unique factor can offset the negative effect of subsidiary autonomy on RKT, interview data cannot exclude the possibility that other positive aspects (e.g., incentives) can also be mitigated. The results of this study represent only the first step of capturing RKT activities inside Chinese MNCs. Future research may complete a paradigm that combines EMNC insights with HQ-subsidiary theory, such as agent theory, to describe the complex process of RKT, which helps in bringing out a more realistic description of reverse knowledge flows implemented across Chinese MNCs.

References

Ambos, T.C., & Ambos, B. (2009). The impact of distance on knowledge transfer effectiveness in multinational corporations. *Journal of International Management*, 15(1), 1–14.

Ambos, T.C., Ambos, B., & Schlegelmilch, B.B., (2006). Learning from foreign subsidiaries: an empirical investigation of headquarters' benefit from reverse knowledge transfers. *International Business Review*, 15(3), 294–312.

Andersson, U., & Forsgren, M. (1996). Subsidiary embeddedness and control in the multinational corporation. *International Business Review*, 5(5), 487–508.

Andersson, U., Forsgren, M., & Holm, U. (2002). The strategic impact of external networks: subsidiary performance and competence development in the multinational corporation. *Strategic Management Journal*, 23(11), 979–996.

Asakawa, K. (2001). Organizational tension in international R&D management: the case of Japanese firms. *Research Policy*, 30(5), 735–757.

Bartlett, C. A., & Ghoshal, S. (1989). *Managing across Borders: The Transnational Solution*. Boston: Harvard Business School Press.

Belizon, M. J., Gunnigle, P., Morley, M., & Lavelle, J. (2014). Subsidiary autonomy over industrial relations in Ireland and Spain. *European Journal of Industrial Relations*, 20(3), 237–254.

Bowman, S., Duncan, J., & Weir, C. (2000). Decision-making autonomy in multinational corporation subsidiaries operating in Scotland. *European Business Review*, 12(3), 129–136.

Buckley, P. J., Clegg, L. J., Cross, A. R., Liu, X., Voss, H., & Zheng, P. (2007). The determinants of Chinese outward foreign direct investment. *Journal of International Business Studies*, 38(4), 499–518.

Cantwell, J.A., & Mudambi, R. (2005). MNE competence-creating subsidiary mandates. *Strategic Management Journal*, 26(12), 1109–1128.

Cavanagh, A., Freeman, S., Kalfadellis, P., & Herbert, K. (2017). Assigned versus assumed: towards a contemporary, detailed understanding of subsidiary autonomy. *International Business Review*, 26(6), 1168–1183.

Chen, W., Ding, Y., Meyer, K.E., Wang, G., & Xin, K. (2018). *Global Expansion: The Chinese Way*, London: LID Publishing.

Chiao, Y., & Ying, K. (2013). Network effect and subsidiary autonomy in multinational corporations: An investigation of Taiwanese subsidiaries. *International Business Review*, 22(4), 652–662.

Child, J., & Rodriguez, S.B. (2005). The internationalization of Chinese Firms: a case for theoretical extension? *Management and Organization Review*, 1(3), 381–410.

Chung, L. (2014). Headquarters' managerial intentionality and reverse transfer of practices. *Management International Review*, 54(2), 225–252.

Ciabuschi, F., Kong, L., & Su, C. (2017). Knowledge sourcing from advanced markets subsidiaries: Political embeddedness and reverse knowledge transfer barriers in emerging market multinationals. *Industrial and Corporate Change*, 26(2), 311–332.

Collinson, S.C., & Wang, R. (2012). The evolution of innovation capability in multinational enterprise subsidiaries: dual network embeddedness and the divergence of subsidiary specialization in Taiwan. *Research Policy*, 41(9), 1501–1518.

Deng, P. (2009). Why do Chinese firms tend to acquire strategic assets in international expansion? *Journal of World Business*, 44(1), 74–84.

Eden, L. (2009). Letter from editor-in-chief: reverse knowledge transfers, culture clashes and going international. *Journal of International Business Studies*, 40(2), 177–180.

Edstrom, A., & Galbraith, J.R. (1977). Transfer of managers as a coordination and control strategy in multinational organizations. *Administrative Science Quarterly*, 22(2), 248–263.

Edwards, R., Ahmad, A., & Moss, S. (2002). Subsidiary autonomy: the case of multinational subsidiaries in Malaysia. *Journal of International Business Studies*, 33(1), 183–191.

Eisenhardt, K.M., & Graebner, M.E. (2007). Theory building from cases: opportunities and challenges. *The Academy of Management Journal*, 50(1), 25–32.

Fenton-O'Creevy, M., Gooderham, P., & Nordhaug, O. (2008). Human resource management in US subsidiaries in Europe and Australia: centralization or autonomy? *Journal of International Business Studies*, 39(1), 151–166.

Foss, N.J., & Pedersen, T. (2002). Transferring knowledge in MNCs: the role of sources of subsidiary knowledge and organization context. *Journal of International Management*, 8(1), 49–67.

Frost, T.S., & Zhou, C., (2005). R&D co-practice and 'reverse' knowledge integration in multinational firms. *Journal of International Business Studies*, 36(6), 676–687.

Gammelgaard, J., Holm, U., & Pedersen, T., (2004). The dilemmas of MNC subsidiary knowledge transfer. In Mahnke, V. & Pedersen, T. (Eds), *Knowledge Flows, Governance and the Multinational Enterprise*, New York: Palgrave Macmillan.

Gammelgaard, J., McDonald, F., Stephan, A., Tuselmann, H., & Dorrenbacher, C. (2012). The impact of increases in subsidiary autonomy and network relationships on performance. *International Business Review*, 21(6), 1158–1172.

Garnier, G.H. (1982). Context and decision-making autonomy in the foreign affiliates of U.S. multinational corporations. *Academy of Management Journal*, 25(4), 893–908.

Gupta, A.K., & Govindarajan, V. (1991). Knowledge flows and the structure of control within multinational corporations. *Academy of Management Review*, 16(4), 768–792.

Gupta, A.K., & Govindarajan, V. (2000). Knowledge flows within multinational corporations. *Strategic Management Journal*, 21(4), 473–496.

Haier official website, (2021a), "Haier Malaysia", available at: https://www.haier.com/my/about-haier/news/20190604_74016.shtml (accessed, 10 June 2021).

Haier official website, (2021b), "Haier Pakistan", available at: https://www.haier.com/pk/about-haier/haier-pk/?spm=pk.28248_pc.footer_86023_20190530.3 (accessed 10 June, 2021)

Jean, R.J., Sinkovics, R.R., & Kim, D., (2010). Drivers and performance outcomes of relationship learning for suppliers in cross-border customer-supplier relationships: the role of communication culture. *Journal of International Marketing*, 18(1), 63–85.

Kanchwala, H. (2021). Haier Refrigerator in India – Review 2021, available at: https://www.bijliba chao.com/refrigerator-fridge/haier-refrigerator-india-review.html (accessed 10 June, 2021)

Kane, G.C., & Alavi, M. (2007). Information technology and organizational learning: an investigation of exploration and exploitation processes. *Organization Science*, 18(5), 796–812.

Konga, L., Ciabuschia, F., & Martína, O.M. (2018). Expatriate managers' relationships and reverse knowledge transfer within emerging market MNCs: the mediating role of subsidiary willingness. *Journal of Business Research*, 93, 216–229.

Kumar, N. (2013). Managing reverse knowledge flow in multinational corporations. *Journal of Knowledge Management*, 17(5), 695–708.

Kumar, V., Gaur, A. S., Zhan, W., & Luo, Y. (2019). Co-evolution of MNCs and local competitors in emerging markets. *International Business Review*, 28(5), 101527.

Liu, Y., & Meyer, K.E. (2020). Boundary spanners, HRM practices, and reverse knowledge transfer: the case of Chinese cross-border acquisitions. *Journal of World Business*, 55(2). DOI: 10.1016/j.jwb.2018.07.007.

Lord, M.D., & Ranft, A.L. (2000). Organizational learning about new international markets: exploring the internal transfer of local market knowledge. *Journal of International Business Studies*, 31(4), 573–589.

Luo, Y., & Bu, J., (2016). How valuable is information and communication technology? A study of emerging economy enterprises. *Journal of World Business*, 51(2), 200–211.

Luo, Y., & Bu, J. (2018). Contextualizing international strategy by emerging market firms: a composition-based approach. *Journal of World Business*, 53(3), 337–55.

Luo, Y., & Tung, R. (2007). International expansion of emerging market enterprises: a springboard perspective. *Journal of International Business Studies*, 38(4), 481–498.

Martinez, J. I., & Jarillo, J.C. (1989). The evolution of research on coordination mechanisms in multinational corporations. *Journal of International Business Studies*, 20(3), 489–514.

Michailova, S., & Mustaffa, Z. (2012). Subsidiary knowledge flows in multinational corporations: Research accomplishments, gaps, and opportunities. *Journal of World Business*, 47(3), 383–396.

Minbaeva, D.B. (2007). Knowledge transfer in multinational corporations. *Management International Review*, 47(4), 567–593.

Minbaeva, D.B., Pedersen, T., Björkman, I., Fey, C., & Park, H. (2003). MNC knowledge transfer, subsidiary absorptive capacity, and HRM. *Journal of International Business Studies*, 34(6), 586–599.

Monteiro, F., & Birkinshaw, J. (2017). The external knowledge sourcing process in multinational corporations. *Strategic Management Journal*, 38(2), 342–362.

Mudambi, R. (1999). MNE internal capital markets and subsidiary strategic independence. *International Business Review*, 8(2), 197–211.

Mudambi, R., Piscitello, L., & Rabbiosi, L. (2014). Reverse knowledge transfer in MNCs: subsidiary innovativeness and entry modes. *Long Range Planning*, 47(1–2), 49–63.

Najafi-Tavani, Z., Giroud, A., & Sinkovics, R. R. (2012). Mediating effects in reverse knowledge transfer processes. *Management International Review*, 52(3), 461–488.

Najafi-Tavani, Z., Zaefarian, G., Naudé, P., & Giroud, A. (2015). Reverse knowledge transfer and subsidiary power. *Industrial Marketing Management*, 48(7), 103–110.

Ndubisi, N. O., Capel, C. M., & Ndubisi, G.C. (2015). Innovation strategy and performance of international technology service ventures: the moderating effect of structural autonomy. *Journal of Service Management*, 26(4), 548–564.

Nobel, R., & Birkinshaw, J. (1998). Innovation in multinational corporations: control and communication patterns in international R&D operations. *Strategic Management Journal*, 19(5), 479–496.

Noorderhaven, N., & Harzing, A.W. (2009). Knowledge-sharing and social interaction within MNCs. *Journal of International Business Studies*, 40(5), 719–741.

O'Donnell, S.W. (2000). Managing foreign subsidiaries: Agents of headquarters, or an interdependent network? *Strategic Management Journal*, 21(5), 525–548.

Persaud, A. (2005). Enhancing synergistic innovative capability in multinational corporations: an empirical investigation. *Journal of Product Innovation Management*, 22(5), 412–429.

Phene, A., & Almeida, P. (2008). Innovation in multinational subsidiaries: the role of knowledge assimilation and subsidiary capabilities. *Journal of International Business Studies*, 39(5), 901–919.

Porter, M.E. (1985). *Competitive Advantage: Creating and Sustaining Superior Performance*, New York: Simon and Schuster.

Rabbiosi, L. (2011). Subsidiary roles and reverse knowledge transfer: An investigation of the effects of coordination mechanisms. *Journal of International Management*, 17(2), 97–113.

Rabbiosi, L., & Santangelo, G.D. (2013). Parent company benefits from reverse knowledge transfer: the role of the liability of newness in MNCs. *Journal of World Business*, 48(1), 160–170.

Rugman, A.M., Verbeke, A., & Yuan, W. (2011). Re-conceptualizing Bartlett and Ghoshal's classification of national subsidiary roles in the multinational enterprise. *Journal of Management Studies*, 48(2), 253–277.

Saliola, F., & Zanfei, A. (2009). Multinational firms, global value chains and the organization of knowledge transfer. *Research Policy*, 38(2), 369–381.

Schulz, M., (2001). The uncertain relevance of newness: Organizational learning and knowledge flows. *Academy of Management Journal*, 44(4), 661–681.

Strauss, A., & Corbin, J. (1998). *Basics of Qualitative Research*. Thousand Oaks, CA: Sage Publications.

Taggart, J.H. (1996). Evolution of multinational strategy: evidence from Scottish manufacturing subsidiaries. *Journal of Marketing Management*, 12(6), 533–549.

Tsai, W. (2001). Knowledge transfer in interorganizational networks: Effects of network position and absorptive capacity on business unit innovation and performance. *Academy of Management Journal*, 44(5), 996–1004.

Tsai, W. (2002), Social structure of 'coopetition' within a multiunit organization: coordination, competition, and intraorganizational knowledge sharing. *Organization Science*, 13(2), 179–190.

Tsai, W., & Ghoshal, S. (1998). Social capital and value creation: the role of intrafirm networks. *Academy of Management Journal*, 41(4), 462–476.

Wang, L., Huo, D., & Motohashi, K. (2019). Coordination mechanisms and overseas knowledge acquisition for Chinese suppliers: The contingent impact of production mode and contractual governance. *Journal of International Management*, 25(2), 100653.

Wang, S. L., Luo, Y., Lu, X., Sun, J., & Maksimov, V. (2014). Autonomy delegation to foreign subsidiaries: An enabling mechanism for emerging-market multinationals. *Journal of International Business Studies*, 45(2), 111–130.

Yang, Q., Mudambi, R., & Meyer, K. E. (2008). Conventional and reverse knowledge flows in multinational corporations. *Journal of Management*, 34(5), 882–902.

Yin, R. K. (1984). *Case Study Research: Design and Methods*, Beverly Hills, CA: Sage.

Young, S., & Tavares, A.T. (2004). Centralization and autonomy: Back to the future. *International Business Review*, 13(2), 215–237.

Zhou, A. J., Fey, C., & Yildiz, E.H. (2020). Fostering integration through HRM practices: an empirical examination of absorptive capacity and knowledge transfer in cross-border M&As. *Journal of World Business*, 55(2), 100947.

Yibing Zhang and Siew-Huat Kong

3 Learning and Innovation in Chinese Firms: The Implication of Quanzhou Managers' Cognition in a Transition Era

Introduction

China was able to attain the status of the "factory of the world" and the world's second largest economy within a few decades were due in small measure to her ability to transform herself under the overarching "open-up and reform" policy launched in 1978 (Tisdell, 2009). This nation-wide transformation over the past 40 years is well illustrated in the city of Quanzhou, which is a typical industry powerhouse with large-scale manufacturing of clothing and footwear products for both domestic and international market (Zhang, 1997) and they are in the private sector.

To begin with, Quanzhou has witnessed several stages of its industrial evolution. In the 1980s, the values that were emphasized were to behave well, to be honest in business, and to be an astute businessman, as cheating would cause market chaos. Most enterprises would take risks in any endeavor, as there was money to be made in every business field. In the 1990s and the 2000s, most enterprises directed their focus towards real estate, e-commerce, finance business, and the Internet. More than 10 years ago, due to the excess supply in the property market, the burden of rising debt, and the fall in both revenue and profit for clothing and footwear, business transformation has become a popular discourse in Quanzhou. Managers have been preoccupying themselves with idea of innovation or transformation. The quest for transformation was given added momentum since Premier Li Keqiang proposed mass entrepreneurship and innovation at the 2014 Summer Davos in Tianjin. In the meantime, China's economic growth rate slowed to a 25-year low of 6.9% in 2015, as the world's second-largest economy continues to shift away from its manufacturing roots (CNBC, 2016). Though they, especially the new generation of entrepreneurs, want to transform the traditional factories of China – Chinese manufacturing – into something else, they don't know what that something is yet.

Due to the transitional nature of the Chinese economy and the dynamic pattern of managers' cognition, it is all the more necessary to explore how managers' cognition are formed and manifested in the pattern of behaviors in Chinese firms. This study endeavors to transcend what the researcher perceives to be the fragmented current approaches to reading managerial thought and behavior. As its working construct it proposes a more integrative approach, the "intellectual framework", taking the ontological assumption that the essence of an organization is a system of thoughts.

https://doi.org/10.1515/9783110715002-003

Using an ethnographic method, the study examines a group of managers residing in areas surrounding Quanzhou that are understood to be going through a period of economic transition. Based on in-depth interviews with 50 managers and participant observation conducted in Quanzhou over a period of over one year, this paper examines the intellectual framework of managers in that city, as a "trial run" in providing a new lens for organizational transformation study.

The themes emerged from this fieldwork contribute to the development of the intellectual framework for transition that can help to account for managers' behavior and practices in such diverse aspects as the use of copycatting, strict internal control system, financial IPO strategies, the fever for incessant movements, or excessive investment.

The study suggests that these managers' cognition about transformation is evolving organically as new sources of knowledge are introduced. As Quanzhou is one among a number of economic powerhouses in China, the findings of this work will also contribute to the understanding of Chinese managers in other regions.

Organizational Change and Managers' Cognition

Organizational Change

Most change models cast change in this manner: unfreezing – change/learning – refreezing (Drucker, 1994; Lewin, 1947; Schein, 2004). It is a process entailing frame breaking, and the formation of a new frame (Zohar, 2004), unfreezing of old traditions and experiences and refreezing with new knowledge. Importantly, the evolution of a frame has a direction: more coherent, effective, clear, comprehensive, elaborate, and insightful.

Drucker (1994) posits that old pieces of knowledge that don't fit the reality should be challenged and adjusted according to the changing environment. A stream of management and organizational research has started to address ways in which individuals may change the pieces of knowledge in the foundation, or construct new cognitive frames altogether, in an attempt to overcome the rigidities of existing frames (Cornelissen & Werner, 2014; Gavetti & Levinthal, 2000). This may involve a comparison between a mental frame that individuals know very well, and another target guiding framework that they are eager to re-exam, and where the analogy provides a potentially new framework, with new sources of knowledge, new insights, inferences, and dimensions. In order to infer an alternative frame, new insights may be obtained through accumulated shared learning (Schein, 2004), personal mastery (Senge, 2006), emergent strategy, or grassroots strategy-making (Mintzberg, 1989).

Managers' Cognition

Corporation: a thought system. The corporation can be conceived as a community that is entirely organized by thought, with emotion, cognition, and social structure linked to this system (Bohm, 1994). By basing itself on certain assumptions of social theories, by changing its material institutional conditions and practices, each corporation would express its own thought style, which is penetrating the minds of its members, defining their experience, and setting the poles of their moral understanding (Douglas, 1986). This study is sympathetic to Bohm's ideas on organization, but attempts to use a new angle in looking at organization and its transformation.

The thought system of the corporation is closely interlinked with the manager's mental model (Senge, 2006). Although people are not always consistent in what they say and do, they do behave congruently with their mental models (Argyris, 1982). Managers see and experience, while infusing their imagination, traditions, and knowledge into what they see.

Managers' cognition and transformation. Their mental model might, however, limit managers and block their creative energy for innovation (Senge, 2006). Although managers are able to access infinite information from the world around them, they are very selective as to the concepts that they would adopt, particular decisions they would make, and pay attention to only particular issues, and formulate particular strategies that determine the main structure of their cognition. Also, their direction of thoughts is about the visible problem rather than being set by an understanding of the forces operating behind the problem. As Einstein (1946) put it, no problem can be solved from the same kind of thought that created it. Therefore, only by coming to understand the patterns that are being formed in their own behavior do they come to know their capabilities and their potential (Mintzberg, 1989; Schein, 2004).

Understandably, the cognition of the business managers is at the core of transformation of Chinese business and hence they would constitute interesting subject of investigation. From time to time, in order to prepare for a change, managers have to adjust those elements in their framework that do not fit reality anymore (Drucker, 1994). What is the state of the managers' cognition in this period of transition and what are the implications of these cognition on Chinese managers are questions that are yet to be explored in full.

Characteristics of Managers' Cognition

Firstly, managers' cognition is harder to change. The cognition is the set of habits and conditions that compel managers to act as they do. These are selected concepts that govern the way they think and act (Forrester, 1971). Through defining the goals, incentives, costs, and feedback, the intellectual framework motivates or constrains managerial behavior (Isaacs, 1999). A cognitive framework is relatively stable. Habits are the result of thousands of years of experience handed down and transmitted from one generation to the next (Weick, 1995). While these habits are initially conscious, they may – over time and through repeated usage – evolve into naturalized practice (Goffman, 1974). Managers tend to repeat a pattern in accordance with their old behavior. Either implicitly or explicitly, consciously or unconsciously, they are consistent in basic ways. Any challenge or questioning of the key elements of this framework will release anxiety and defensiveness. In fact, if the fundamental elements of this framework come to be strongly formed and shaped in group, members will find behavior based on any other premise inconceivable (Schein, 2004).

Secondly, cognition is growing organically. Though this framework is often known to demonstrate its resistance to change, it is not static; on the contrary, it is something that is dynamic and is treading its own path of evolution all the times.

Thought systems exist and constantly engage in a process of development, change, and evolution (Bohm, 1994). Zohar (2004) argued that human consciousness itself displays the characteristics of complexity in many of its abilities, like complex adaptive systems. The way of thinking seems to be changing constantly, especially for those who are more open-minded, humble, and eager to learn new things. They are more receptive to the new knowledge, new standards, new capacities, new experience, and the diversity of opinions (Wheatley, 2006). Therefore, managers' cognition has infinite degrees of perfections.

McGregor (1960) explained that through the power of rewards and punishments, and a learning process, there will be changes in attitudes, perceptions, and behavior. These learning opportunities include ones that are external and tangible, for example, praise from the boss; or that are internal and intangible, for example, the frustration of being blocked in pursuing one's goals. A manager's experience, reflection, and learning from mistakes might all be forces that enable this framework to grow and develop. Therefore, the cognitive domain, which helps shape the manager's behavioral domain, is consistently changing as a result of ongoing interaction between individual and organizational dimensions, between overt and covert, and between inner world and outside.

Thirdly, managers' cognition is tempered by the environment. Bohm (1994) also states that thought structures are primarily collective phenomena. There is a mutual interactive relationship between the thought of the individual and that of the collective. Some would propose that the social nature inherent in our cognitive activities ought to be given due consideration. Fleck (1981) argues that "without social conditioning no cognition is even possible. Indeed, the very word "cognition" acquires meaning only in connection with a thought collective." (p. 43) "Because the entire fund of knowledge as well as intellectual interaction within the collective take part in every single act of cognition, which is indeed fundamentally impossible without them" (Fleck, 1981, p.43). This position then is different from the prevailing premise that human intellectual activity is one that an individual carries out on his own, and is independent from the environment or the collective that he is in. While I do not take Fleck's position in its totality, and I readily subscribe to the position that an individual is capable of generating independent thoughts, but I am more comfortable to say that the degree of independence of our thought is tempered by the environment or the collective that we are in.

By building more elaborate models, individuals and groups are embedded within institutional contexts, where expectations are structured and pattern of behavior is scripted. Literature on the strategic framework of change focuses precisely on the linking between framing in communication and the frames of understanding or explanation of all staff members in an organization (Gilbert, 2006; Huff, 1990; Kaplan, 2008; Nadkarni & Narayanan, 2007). Within this strand of literature, the construct of cognitive framework is generally referred to the intentional communication efforts of key management team in shaping the frames of perception of other members in an organization, so that they, as a group, would support a change program (e.g., Bartunek, 1993; Garvin & Roberto, 2005; Gioia & Chittipeddi, 1991; Kotter, 1996; Mantere et al., 2012).

Transition: China and Quanzhou

China has been implementing policies which promotes "Open Door" and reform since 1978. After many years of "experiment", the nation appears to be once again at a cross road and the government is hoping to deepen the effect of its policies in order to address those challenges.

In today's fast-changing environment, the clash of traditional and modern value systems, the Chinese and Western cultures, and the various discourses on the common problems faced by human being is a fertile ground in diagnosing the basic attitudes and underlying assumptions of a society. The assumptions, beliefs, views, values and approaches of Chinese managers have emerged from a specific cultural tradition (Redding, 1993; Witzel, 2012). Redding (1993) argues that certain ideas derived from Chinese cultural tradition, such as paternalism, face, *guanxi* (interpersonal relationships), filial piety, and pragmatism, have played a significant part in determining the economic behavior displayed by those managers.

Since 1978, Chinese society has been in direct contact with foreign concepts, cultures, technologies and lifestyles as a result of reform and the open-up policy. Globalization, foreign direct investment, and the pervading of the Internet have exposed China to unprecedented global knowledge transfer, information sharing, and culture learning (Faure and Fang, 2008). During this transitional stage, due to the lack of a clear and agreed upon framework of thoughts that would help them to orient their behavior, many Chinese people are anxious about their future lives, and some felt that they have even lost touch with their inner selves (Osburg, 2013; Reed, 1991; Madsen, 2011; Vernezze, 2011). Osburg (2013) points out that such anxiety calls for the adoption of new value systems, which would prevent human relations from being contaminated by market forces, and protect human personalities from being shaped principally by monetary value.

Like other areas or regions on mainland China, Quanzhou is undergoing this process of transition. The traditional industries in this region are encountering problems that have been confusing the managers there. They need a deeper understanding of their current practices, and especially those pieces of knowledge in their thought structure that are creating the problem, which is certainly not able to provide a remedy for their firms' transformation at the present.

Reasons for this Study

This study is to investigate the influence of cognition of managers on enterprise management, vis-a-vis transformation. It aims to demonstrate the potency of managers' cognition towards the future development of their business, which permeate all facets of organizational and individual life despite how little people actually are conscious of it. In other words, the aim is to show the expression of cognition in the context of Chinese organizations, which are undergoing transformation themselves.

The reason for exploring the cognition of the Chinese managers is simply to obtain a better understanding of their managerial behavior and its connection to its ultimate sources. It is commonly accepted that managers, organizations, and managerial practices are all potentially influenced by traditions, theories, and conventions. By conducting fieldwork in a region, there is a chance to explore these sources and to see their influence on how managers see and act toward the external world. In other words, managers must understand where the new sources of thought come from, in the context of ongoing and deepening reform in a new China of today. If the Chinese managers talk about change, this study seeks to identify their current position, as well as where they are trying to go. Admittedly, intellectuality, consciousness, and thought are abstract terms, but the attempt of this study is to describe their related elements, attributes, and characteristics in a more concrete and understandable way, and hopefully provide a new angle to study management and organizations in the context of transition.

Research Questions

The present study is based on the premises that cognition is the most socially-conditioned activity of man and the management system of a human organization is animated and sustained by the collective thoughts that permeate that entity (Fleck 1981). As such, this study will focus on the cognitive aspect of management

(Argyris 1976; McGregor 1985; Schein 2004; Senge 2006; Weick 1995). Specifically, this investigation addresses the following questions:
- What are the different elements of managers' conceptions about transformation?
- How did the conceptions about transformation come into being, and how are they giving shape to the current structure of management and managers' behaviour?

Research Methods and Setting

For ease of narrative, the fieldwork is described here from the perspective of one of this article's contributors. I made contact with the informants via classmates, ex-colleagues, friends, and family members. The informants (see Appendix) are from a wide variety of backgrounds, although the majority of them hold positions as middle or senior management in private firms. These firms are of different sizes, from less than one hundred to several hundred employees, and in such diverse industries as garments, shoes, or leather manufacturing; cultural and creative products; and real estate development and sales.

Information Collection and Analysis

There were in total around 250 hours of in-depth interviews and participant observation. As well as data collected from interviews and observation, some sources of documentary evidence were also considered, such as companies' annual reports, employee manuals, company policies, mission and values statements, company magazines, official websites, news outlets, and newspaper reports. Most of the interviews were recorded. Each of the fifty interviews was transcribed into a separate Word file. Following Agar (1996), a summary of the key points regarding observations, conversations, interpretations, and suggestions for future information was drawn up. The final written language of transcription was Chinese, both Mandarin and the *Minnan* dialect. It took around 240 hours to complete the transcription of the fifty interviews. The result of initial coding provided thematic analysis of the interview transcript using nodes. Software was used to help in the process of data indexing, managing, and interpreting. The nodes were examined by categorizing them.

In order to improve this study's validity, prolonging the engagement in interviews and observation was considered. Triangulation through using ethnographic interviews, informal interviews, stories, document analysis, and participant observations was used to reduce misunderstanding and distortion. Also, showing field notes to informants for checking the correctness of data collected was also a vital part of this fieldwork.

Findings

According to the majority of informants' responses and my observation, the main issue for Quanzhou managers is how they are going to transform from the traditional industries to the so-called modern industries – although it is not clear what these might consist of. Some managers have the intention of getting out of manufacturing, and moving to something else. Others would like to stay in the industry, but the compulsion to engage in transformation is more than real for them. There are concerns about the unprofitability of traditional industries and many have questions about the role of innovation in such transformation. My findings will also reveal the conflicts between new ideas and old ones regarding transformation, which is more clearly demonstrated between the first generation and the second generation entrepreneurs, or between the parents and their children. The findings of the study are discussed next, grouped around several themes, as below.

Innovating or Maintaining Status Quo

> *Innovation is difficult, but it is more difficult without innovation; transformation is exhausting, but it is more exhausting without transformation* – Lin Jinben, Chairman of Fujian Energy Group Corp. and one of 30 Fujian entrepreneurs who wrote President Xi Jinping a letter in May 2014

There is a lot of anxieties and concerns on the part of entrepreneurs about China's economic condition. Questions have been raised regarding the need for deeper economic reform, such as: "What is the biggest risk facing China?" and "Are reforms needed in order to keep the economy going forward?" Much transformation has in fact taken place in the past several decades – 1978 was the year of farmer entrepreneurship; 1984 is known as the first year of enterprise; in 1992, the government encouraged the masses to jump into the sea (of commerce); 1998 was the Internet revolution; and the current decade is about mass entrepreneurship and innovation. Undoubtedly, the Quanzhou entrepreneurs were also trying hard just to catch up. Many informants pointed out that China has become the world's factory, focusing on less value-adding aspects such as production and manufacturing; moving away gradually from this scheme of economy would require nothing less than innovation.

Copycatting

The clothing and footwear industry in Quanzhou is a mature one. Started in the 1980s in Shishi, a Quanzhou town, the industry ably supplied the domestic market. After the 1990s, competition put much pressure on the Quanzhou clothing industry, as Zhejiang and Guangdong were rising rapidly. Jinjiang is another emerging industrial town in Quanzhou. At the beginning, it was a little-known rural town in the coastal province of Fujian. Gradually, it became the biggest export center for sports shoes. One of my informants, Chen Chang, told me:

> Quanzhou, this new industrial city, is a major center for making clothes, zippers, and toys. There are thousands of shoemakers and dozens of nationally known brands. These shoemakers employ several hundred thousand workers and produce more than 500 million pairs annually.

Shishi and Jinjiang are famous for *shanzhai* (copycatting). When somebody has worked very hard for half a year to design a pair of shoes, it only takes four hours for thousands of shoemakers to copy that design. Liu Dong said:

> An extraordinary array of talented young people is busy with copying others' design online, rather than producing original designs. Thus, we cannot find an enterprise designing fashionable clothing anywhere across the globe. Currently, we don't have stunning design. The clothing business has been so depressed and foreign brands are taking advantage of the situation. Copying is an important cause for the defeat of the whole industry.

According to a survey conducted by the website tech.qq.com, nearly 95% of Chinese Internet users believe that counterfeit goods are "running wild" online (Ma, 2015). Some studies argued that this is a necessary stage for China to go through, because the USA had has 70-year history of 'copycatting', Europe 30 years, and Japan 15 years before attaining their industrialization status (Ma, 2015).

As of 2020, The 'Made in China' label is still associated with low quality and low price. The criticism has been made that there are no major national enterprises like Samsung in Korea or Sony in Japan. There are a lot of low-level and redundant projects. The sceptics insist that most companies will never be able to show good performance, because they will always be tempted to cut corners in order to maximize their profits. Some informants take the view that once early adopters latch onto something new and useful, the rest of the firms would quickly follow.

Financial difficulties. The majority of informants are concerned about the economic recession and difficulties faced by enterprises. Some firms plunge into financial distress in the fierce market. When everybody was involved in the clothing business, the company faced the capital chain rupture. The suppliers only sell raw material when they have been paid in cash. For the family business, the homogenization crisis was a painful memory. Many of them finally collapsed. Lin Shui said:

> Copycatting operates so efficiently in clothing and footwear. Big fish eats small fish. Sixty percent of enterprises project a deficit due to severe competition. Enterprises are very weak in anti-risk capabilities. Every day I hear that enterprises around us have collapsed. There are only a few brand names that still have hope, like Anta and Hengan. Other factories dare not sell their products online due to low quality. I have been running my business for nearly 38 years. Today is extremely difficult. In the 1990s, I could easily borrow 100 million yuan from my acquaintances. But now borrowing 5 million yuan is impossible.

Many factory owners can't see when the market will experience a rebound so they are trying to cut their losses by winding up their business, before all their money is gone. I knew a driver during my fieldwork in Jinjiang, Ding Shen, who was a former footwear factory owner. One day he chauffeured us to the airport. On the way he told me:

> I used to owne a shoe factory in Jinjiang. I had been running the business for more than ten years. My business was not operating so well. Especially, shoemakers were competing with each other and I couldn't make money anymore. So I decided to shut my factory down and became a full-time driver.

I think his factory must be among thousands of factories that were shut for good. Due to the difficulties some manufacturers faced, Quanzhou local government is undertaking financial reform, encouraging more private resources to be directed to this industry. However, some enterprises still cannot get a bank loan, as an estimated 70 percent of loans were issued to state-owned companies (Zhao & Arvanitis, 2010). It is reported that 18 enterprises including Nuoqi were reported to have "runaway bosses", a phenomenon that is becoming a 'new normal'. Also, there are more than 70 enterprises in Quanzhou that are reported to be lacking in finance creditability.

CCTV advertisement. In order to sustain their business, many owners attempt to take higher risks. Lu Hai Said:

> Quanzhou merchants are interested in gambling. The typical example is mark-six lottery. Quanzhou is one of the most famous cities in China for this game. Many people lose their money in such gambling. They are eager to become successful right away. In the 1990s, many normal families, bosses put all of their money into it.

Risking one's whole fortune in CCTV advertisement is another example of gambling. It is reported that Ding Shizhong invented the business model of advertising through sports star and CCTV. In 1990s he invested several million yuan, equating to years of profit or running the family business, in advertising. Subsequently many bosses copied this model. Before long, CCTV 5 became the Jinjiang channel. For the new generation, these events are exciting adventure, and unnerving.

The company Lilang undertook this kind of gambling in 2002. Wang Liangxing, the company's CEO, borrowed several million yuan to pay for certain movie star Chen Daoming's advertising fee. At that time, his business had a liability of about 30 million yuan. He also got money from friends. "The victor is the king, the loser is the bandit," Wang said. In 2002, the company's revenue was only 40 million yuan. In the end he won the battle. In 2004, Lilang's revenue became 100 million yuan. Then, Wang made a bigger bet. He raised money from friends and decided to pay 10 million yuan for advertisement for 16 days during the Olympics. That year the revenue of Lilang increased by 400%. In reality, though, Lilang was short of cash.

The phenomenon of the CCTV advertisement reveals the preference of Quanzhou merchants for taking risk, and their belief in the effect of a movement. "Make the most of a bad situation" refers to the attitude of taking risks. And if they should ever succeed, they will become rich overnight, many like to believe.

Taking risk seems to be the main character of Fujian entrepreneurs. In May 2014, 30 Fujian entrepreneurs wrote President Xi Jinping a letter on the subject: "dare to be responsible, risk to be excellent." The purpose is to express the intention of speedily reforming and developing enterprise. In his reply to the entrepreneurs, President Xi Jinping emphasized the importance of creating a better environment for enterprises development all over the country; all levels of government are required to speed up the change of their functions and encourage simplified administrative procedure. He also mentioned better and orderly development of the market, a new opportunity for entrepreneurs to create an innovative future, a role for the market in resource allocation, and building a competitive and fair environment. The tone, if not the spirit, of the letter certainly resonates with the "Quanzhou spirit" animating Quanzhou entrepreneurs in the past 30 years: "Dare to be a forerunner, Dare to fight for the prize".

Transform to what. When I asked my informants, "transform from what to what?", many of them had no answers. Kang Jing told me that his company is selling projects at a cheaper price. Some enterprises are taking steps to diversify their business. Many firms still have the purpose of expansion and increasing productivity. Lin Shui told me:

> *We are investing in more machines currently. I have to take this risk, as I have no other means to ensure the survival of my business. It is just like gambling.*

Wu Ji, the company's Financial Controller, said:

> *My boss is always making the wrong decisions. This is the reason why we are losing more and more money. He is enthusiastic about investment, although he is more cautious in recent years. I think most bosses have considered transition from traditional businesses to new ones, but they don't know how.*

Lu Hai's boss was born in 1972. He established his business some 15 years ago. In the last five years, due to the continuing recession in the clothing industry, he is finding it difficult for his firm to get breakthroughs in its business. Other informants echoed the same anxiety and emphasized the importance of producing high quality products. Liu Dong told me:

> *Only if entrepreneurs who prioritize creating better value for clients and society would produce high quality and innovative products. Why did Nokia lose market share and was pushed out of the international market by Apple? The reason is that Apple is able to create better value for customers. Most private businesses in Quanzhou are manufacturing based. They have a rich industrial foundation. If entrepreneurs have determination to change, they will have a lot of room to develop within the traditional industry.*

There are many problems and challenges faced by traditional industries. Homogenization of product and production process are serious among enterprises. When a product is selling well, many people just jumped on the wagon and the market become saturated with too many player very quickly. The test for many enterprises now is coming up with product of high quality. Actually, many Quanzhou businesses started in the early 1980s, which also witnessed the establishment of some famous brands such as Anta and Tebu. Today, the good old day is behind them, and to stay on the same track is to impose a death sentence upon themselves.

Most of the enterprises in Quanzhou seem to be experiencing a difficult time. There are new challenges for manufacturers, especially after they have gotten used to one way of doing business and don't have the motivation for innovating. In traditional manufacturing, the design and production procedures are predetermined, with the design and production processes that can be copied quite easily. Unfortunately, the same rules do not apply any more these days.

For my informants, the new ways of thinking, especially those that are related to the Internet, are just too overwhelming for them. Some of these companies are turning to the consulting companies for assistance. Also, enterprises like Hengan, Anta, and Qipilang are some of the role models in their environment for them to learn.

Management Transformation and Internet Plus

As mentioned by my informant Zheng Xin, renowned enterprises in Quanzhou like Qipilang, Anta, Hengan, and Lilang stay a step ahead in terms of reform, especially in the realm of brand strategy. For example, Qipilang is in its third stage of managerial reform, deepening its brand name through quality improvement at an unchanged product price. Qipilang has been raising consciousness among senior management that innovation is the only way for sustaining the firm. In order to strengthen brand marketing and to establish brand awareness, Qipilang diversified its portfolio of business models, brands, and geographies. Lu Hai told me:

> Qipilang is our benchmark enterprise. It was established in the 1980s and had very good revenue for several years. At that time, it was very easy to get money from banks. They bought many properties and high-end machines. They launched large-scale investment and expansion. However, banks always close the umbrella when it is raining. They could not get money from banks anymore, as China aimed to strengthen macroeconomic control and continued with a tight monetary policy to prevent its economy from overheating.

> Afterwards, they tried to adopt a new marketing strategy, first attempting the franchise model. Then, the brand name expanded to alcohol, tobacco, tea, and watches. Again in 2013, the company faced a drop in revenue and profit.

> The core value of Qipilang is to remain in a state of struggle, like the character of the wolf. However, as the company has made remarkable achievements in recent years and the wolf has enough food, it is not so aggressive anymore. In addition, the economy is declining. Wanggou

(shopping online) is eating the cake of retailers. Also, it is not easy to raise money from the non-banking sector continue to drop.

Therefore, recently the founder invited all partners to start a new undertaking, encouraging employees and partners to recall the wolf spirit. Qipilang interpreted its brand as "man has more than one face." In 2010 it developed "mingshitang," inviting movie stars like Sun Honglei and Hujun to get involved in marketing wars. It aimed to present a different style of movie star, but with the same aggressive character, to build a new Qipilang brand.

Certainly, it is a very successful enterprise, but they have to innovate and reform continuously. What we can do is just study what they are doing, and follow them.

Anta, being a role model in the eyes of my informants, is a company frequently referred to by the informants. A cursory review of its revenue over the course of the company's history shows that, starting in 1991, it took the firm ten years to achieve 1 billion yuan in sales revenue. Then, it spent another ten years to achieve the revenue goal of 10 billion yuan. The company plans to take the opportunity of the next ten years to achieve 100 billion yuan in revenue. In his address to employees in 2015, Ding Shizhong, the CEO of Anta, said: 'What we are doing is subversive. But we need to be in the state of struggling, meaningful struggling. We are struggling for efficiency and competitiveness. We subvert ourselves so as not to be subverted by competitors' (Huaxiajishi, 2016).

He quoted Leijun (the founder of Xiaomi) as saying: 'In this age, everywhere is in the teeth of the storm. We are at the forefront, where the pig is able to fly.' He pointed out that China's economy is experiencing a new round of transformation and upgrading, the company should be able to achieve a revenue of 100 billion yuan over the next ten years when it is riding on the latest wave of national transformation. However, the major challenge for the company is that the global economy is presently facing huge difficulties and most businessmen are in a bad mood, and are fickle.

However, the vision for Anta is clear: to be No. 1 in the world sports utility business. As Ding Shizhong says, they do not aim to become China's Adidas, they want to become a global Anta. In order to strengthen its brand name, Anta needs to have a lot of funds. One of Ding's major concerns is how to raise the capital required to realize such a vision. Ye Xiang explained to me:

Anta, in order to achieve its commercial ambition, has to conduct reform, especially in its brand strategy. With such an ambition, Anta needs money and resources, to support its reform and transform. Ding Shizhong is rendadaibiao (the National People's Congress delegate). He has submitted several bills, suggestions and recommendations, all related to the need for financial support and internationalization of private enterprises.

Xu Lianjie, who is regarded as the godfather of Quanzhou merchants, continuously leads his enterprise Hengan to transform and reform. Some informants told me that many successful local business owners were always seeking guidance from Xu

Lianjie. In turn, Hengan has been relying on international management consulting companies for the necessary inspiration and expertise.

In 1996, Hengan started to operate its 'Anerle' brand baby diaper business. However, sales were only floating at 1 billion level for several years. In 2009, in order to drive its 'second-wave' management reform initiative, Hengan initiated full-scale implementation of its Performance Excellence System. It set a goal of sales exceeding the benchmark of 10 billion. In 2014, Hengan and IBM signed a project contract to start 'third-wave' reform, which aimed at establishing Hengan's big data industry pattern. Today, Hengan has a capitalization of 100 billion. Xu believes that business has to seek support from big data. Many informants agree that in order to keep up with the changing market, Hengan places high emphasis on buying services from consulting companies. Lu Ying told me:

> More and more private firms have realized that managing in a crude way cannot sustain the growth of business anymore. Xu led the enterprise going public in HKex in 1998. He sees this opportunity for building a modern enterprise system. Hengan is among a few companies that has benefitted from the process of IPO. IPO is the bedrock for Hengan's future development, enabling it to keep up with the latest market developments in the world, and maintain competitiveness in the tissue market.

From the responses of my informants, however, most firms' transformation seems centered on production and sales and marketing, rather than designing new products or services. The majority of family owners have been hesitating about management transformation. In fact, the old mode of management cannot be changed in a short time. Since the overarching goal for innovation is profit making in the short term, the reform for most firms has been superficial.

Internet Plus. Though the first-generation managers are concerned about the distraction of time and energy accorded to unfamiliar business areas such as the Internet, the Internet seems to be at the core of innovation, especially for young business leaders. Studies (Caijinglangyan, 2016) show that China's traditional enterprises use the Internet as a means to reach out to clients at a rate of 25%, a very low percentage compared with America's 75%. "Internet-related" economy reached 35% of the GDP in 2020. As China's economy is showing sign of slower growth, the government is trying to create a new driver to stimulate the country's development. Hence, "Internet Plus" is named as a national strategy in the Government Work Report of 2016. Some Chinese entrepreneurs in the IT industry place great emphasis on Internet Plus. Liu Qiangdong, the Chairman of JD Group, said: "In the past twenty years, the Internet has been separated from offline businesses. However, in the next twenty years, they will share the same breath." He pointed out that the Internet will be a much more robust force in China because offline businesses are currently much less efficient.

Recently, Premier Li Keqiang made remarks during his inspection of two Quanzhou enterprises. He stressed the importance of Internet Plus. Due to rapid economic development and the improvement of the mass living standard, some informants are impressed by the service industry flourishing in some Chinese cities. Tmall, Taobao, and Jingdong are at the center of the battle for the e-commerce market, currently experiencing phenomenal growth. Internet sales seems to be everything, and online shopping fever is not showing any sign of decline. At the same time, courier companies have rapidly expanded. However, this kind of exciting development has been focusing on promotion and distribution of products in the domestic market.

Some informants think e-commerce is stealing market share from the traditional industries, and is doing so very rapidly. The recession is taking a serious toll on manufacturing, and e-commerce could emerge as a winner. E-commerce was often emphasized by my informants Lu Hai, Ding Hua, and Chen Qi during conversations. Ding Hua is directing his efforts towards the strategy of combining traditional ecommerce with the new model of e-commerce, a B2B business model. He contacted me on a recent visit he made to Macau; over dinner, he asked me about a well-known person in Macau who is a knowledgeable expert in the field of big data. Ding Hua's enterprise is a traditional shoe manufacturer in Quanzhou. For 30 years the business operated in labor-intensive cloth shoes business. Ding and his brother-in-law are working on a new brand strategy. The brand name 'Dengzu' was developed in the wholesale market. After 2005, it had breakthroughs in both production capacity and sales. As a production-oriented enterprise, with sufficient cash, the consideration will be building its own brand name. For Ding Hua, the road map for the present reform is clear: it is the Internet.

Ding Hua is among the children of entrepreneurs I interviewed. He is most interested in the new market arising from the application of the Internet. Chen Qi, another rich second generation, shared similar passion. Among my respondents, Ding Hua, Wu Mei, and Chen Qi are the second-generation entrepreneurs who do not tread the path of their parents in terms of mind-set. This appears to be the norm for the family businesses in this city, as the founders' children tend to have different kinds of education and life experience as compared to their forebears. Their pre-occupation is more about how to manage the capital, rather than the factory, which their forebears did. They are more pragmatic and negotiate on the basis of equality with employees, partners, clients, or suppliers. In their eyes, *renqing* (human relations) is an obstacle for transformation into a modern enterprise. Ding Hua said:

> *Most of the enterprises in Jinjiang are private firms. The leadership succession is vital to the future development of the firms. If there is a successful succession from among the family member, the future of the firm might be secured. The other option is going public, introducing a professional management team. For a family firm, this is the end of the story. Few enterprises I know in Jinjiang have a successful succession. The fate of an enterprise is like a person. Once the founder becomes suddenly sick and passes away, the enterprise declines too.*

> *My management concepts are from three sources. First, I learn from my father. I am impressed by his diligence, punctuality, and truthfulness. Second, I learn from my father-in-law, the founder*

of the current business. I was influenced by his ideas and mindset at the beginning. But now I don't think his concepts is effective anymore. Therefore, I learn from individuals and organizations beyond our family circle. This is the third source of my knowledge. I am also interested in reading books. I have attended many management courses. For example, I attended some classes at the Chinese-European Business School. I feel I need to listen to different opinions and acquire more knowledge. If I rely only on the old generation's experience and knowledge, I would not be able to sustain the enterprise.

Obviously, this new generation is not embracing the traditional footwear and clothing business. In some ways this situation is very similar to a lot of other cities, especially those in the coastal areas. They want to transform the traditional factories of China – Chinese manufacturing – into something else, but they don't know what that something is yet.

New ideas. Nevertheless, some of the rich second generation are quite aggressive. Popular topics among this younger generation include brand strategy, the Internet, innovation, joint venture, capital market, and organizational change. They create their own offspring companies. Some companies are experimenting with Internet Plus, downsizing the old factory. They always have new ideas. They are particularly enthusiastic about the Internet, and its relation to the transformation of the Chinese economy. Ding Hua and Chen Qi both take the view that the traditional industries such as clothing and footwear are too labor-intensive and cannot be sustained in the context of the new economy. These labour-intensive industries will disappear one day, or so they believe. They want to try new businesses that are created around new media and the Internet in their future development. Chen Qi said:

Regarding learning new things, it is not enough by just limiting myself to my father's framework. Some of those principles are good, but my father doesn't want to get out of this framework. Because my father believes that these principles had brought him success and there is no reason for changing it. However, for me who is interested in combining technology with the traditional industry, what my parents have to offer clearly would not do the job.

Headquarter economy. In order to attract more talent, businesses are moving their headquarters to Xiamen. The headquarter transfer strategies give rise to a phenomenon of known locally as "headquarters economy." In the past two decades, more and more firms have started to use the terms HRM. Also, young managers realize that innovation is more vital for the sustainability of a firm.

Anta relocated its headquarters to Xiamen, aiming to attract talented people from many different provinces and those outside of China, rather than just limiting themselves to family members. They are also offering a compensation package that is much higher than other enterprises in the same industry (Zhang, 2016). With more private companies going public, some of the owners are turning their focus to the governance issue. With that, the professional managers are assuming more important role in the management of the listed companies. Liu You told me:

Moving to Xiamen is my dream. My salaries and other benefits will also be improved. The company will benefit from the wealth of expertise and connections I would have over there. For example, as an important part of financial management, my expertise in tax planning can inevitably provide help to the enterprise's financial aspect and even its entire development.

Some of the informants complained that their parents were not keeping up with the trends in human resources management, because they just cling to the old-fashioned concepts or traditions. Wu Mei explained to me:

I went to Xiamen University to attend a course and got to know one of the HR teachers. He has some rich experience in HR, having worked in the enterprise for many years. We shared many thoughts in common regarding people management.

I think decentralization and empowerment are very important. For future management in emerging industry like the Internet, the traditional approaches are not applicable anymore. The old generation might have been successful but their mindsets have not been changing with time and the environment. So, we have several young people within the family forming core groups, committed to transforming our company with the Internet.

Most of the children of the entrepreneurs are the only child in the family. They have had a higher standard of living as compared to their parents when the latter were youth. Most of them have studied or travelled abroad. The values they pursue are more akin to the spirit of capitalism and the wealth myths of the Wall Street. Enterprises like Anta, Qipilang, and Hengan went public at the very early stage. They collected a lot of money and became popular among their peer enterprises in that regard. Admittedly, the rags-to-riches stories had made many young entrepreneurs losing their interest in running a less profitable manufacturing company. They are just too impatient to cope with the slow process related to manufacturing industry; only quick success and instantaneous benefits can please them. Dedication to IPO rather than operating an enterprise has become the mainstream preoccupation for the rich second generation.

Summary of Findings

Succession

There is an old mindset about succeeding under old manufacturing-based business models. Some old entrepreneurs went through the Cultural Revolution and experienced the time of seeing transformation of the economy. They are thankful for the situation now, which is so much better than what they were used to. They know they have to change but the direction and destination are not clear to them yet at this time. They focus more on *renqing*. Their children always criticize them as lacking management knowledge. To be sure, older Chinese managers are only armed with knowledge about production, manufacturing, and engineering. They seem to have expertise only in these areas, and have less knowledge about modern management theories (Warner, 2014). Although they have accumulated knowledge from working in the field, they were not that familiar with the latest theories and practice in areas such as marketing, human resources, government, accounting, and corporate finance, which was made available only several years ago when business schools became popular in China.

In my fieldwork, I discovered that more and more rich second generation is establishing new businesses rather than taking over the businesses from their parents. This is a big shift in family business succession, as traditionally the successor of a family business in China is usually the owner's son or daughter. The traditional way of succession is embedded in the mind of older entrepreneurs, who are influenced greatly by Chinese hierarchical culture. Over the course of Chinese history, filial piety lies at the heart of the *wulun* system, requiring children to obey parents. The Confucian tradition stresses that man exists through his relationships to others; that these relationships are hierarchical in nature; and that social harmony rests upon honoring the obligations they entail (Tang & Ward, 2003). It is thought that people, especially the older generation, are not educated to question what is presented to them. To question somebody or something is perceived to be offensive. People are used to following the rules.

It is generally thought that for the older entrepreneurs to change their way of thinking is a difficult thing. Hence, they tend to protect their industries. For the parents, they still want their children to obey them and to sustain the family business. In some other cities, some children have indeed adopted this value. For example, the successor of FORTILE, Mao Zhongqun, gave up the opportunity of going abroad for a PhD program, to take over the business from his father. Now he is implementing Confucian values in the company.

By contrast, some Chinese new-generation managers who are educated by the Western education system in recent years are more inclined to demonstrate in their behaviour the doctrine of neoliberalism (Harvey, 2005; Wang, 2003). They place more emphasis on self-actualization and tend to promote themselves in every aspect. In a contest culture, they are self-centered and adopt a more aggressive way of management. In this sense, some children of entrepreneurs are beginning to break away from the traditional management mode held by their parents. It seems they are more concerned about transformation, though the direction of transformation has yet to be clarified.

Avoid making mistake. Also, in order to avoid making mistake, some would choose not to take action. Some professional managers would choose to waste their time in the company without any intention of taking action. They believe that if they do something wrong they will have to take full responsibility. However, younger generation managers tend to think that they would not be able to survive in a competing world if they don't take any action all. Some firms adopt the strategy of improving product quality without the price hike. Smiling service, sales and marketing reform, flat organizational structure, technology development, supply chain reengineering, and the internationalization of talents are among their innovative ideas.

Increasingly, the younger generation seem far more willing to try new ways of transformation. They have the energy and are willing to seek new opportunities, and have the ambition of transforming the business by applying knowledge acquired from university or new technology. To implement the strategy of Internet plus manufacturing, initial public offering, or even to move to e-commerce and give up the traditional business are some of the new trends in China for young entrepreneurs. Some managers take the view that the low price and high quality of traditional industries is in itself a way of innovation. Most of them are learning from the experience of certain innovative Western firms. But it seems only a few companies, those with a culture of innovation or inspired by the founders, are positioned to take initiative in these ways.

Attitude towards clients. Another obstacle for transformation is certain entrenched attitude of the managers towards their clients. Some entrepreneur informants complain that the slogan "the user is God" cannot be appropriately applied in their daily management. Since there is no such thing as God, who have ultimate sovereignty over everything, in their belief system, the force for customer-driven innovation is not very strong; owners seldom take the time to know the real needs of their customers. Though they are more than fluent in reciting the rhetoric that understanding customer satisfaction fully is the precondition for the company's development, and they will go all the ways to satisfy customers' needs, and "quality is life, the user is God", in reality they are no more than empty slogan, which are seldom put into action, if at all.

Habit of challenging. Among the older generation of entrepreneurs, not having the habit of challenging their own established pattern of thinking and action also blocks their creative energy. Creativity is often called "thinking outside the box." It is "thinking about thinking." It can take many forms, including knowledge about when and how to use particular strategies for learning or for problem solving (Metcalfe & Shimamura, 1994). When people assert that there is only one right way – experience, tradition, practices, educational training, or old patterns of thinking – and become dogmatic about, as exhibited by some of the older generation, creativity can never be fostered.

Profit making. Transformation and innovation, especially in product that can address the genuine needs of the clients, are not something my entrepreneur informants really love to do; they are just the necessary evils in the management of firms these days. More often than not, they are just some cosmetic measures designed to meet the demand of the fast-changing market, to obey the regulations and policy imposed by government, or to maximize shareholders' value. Though the discourse on reviving the spirit of craftsmanship (工匠精神) is in ascendance, this is not the foundation for competitive strength that my informants are working on. The mentality of taking shortcuts and attaining instantaneous success is an obstacle to innovation for many managers.

For my informants, the motivation for transformation is again centered on quick profit. They complained that they have no other choice because the traditional businesses are facing difficulties of raw material price increases, rising labor cost, decreased product prices, and the inflated value of the RMB. The traditional businesses are operating at an unprofitable level. Further, most manufacturers cannot expand their market in the face of fierce competition. Many managers believe that even if they were to increase their production capacity and investment to enhance their market share, they are not confident that it would succeed. It appeared that the survival mentality or the philosophical idea of *homo economicus* had laid siege on them. For both the older generation and young entrepreneurs, the willingness to undertake transformation is mainly driven by the paradigm of growing the business to a large scale and then going public.

In addition, since the prevailing norm in their business environment focuses more on earning quick profit, firms are not inclined to invest in basic research to develop their competency that is required for them to succeed in their industry. Their present pre-occupation is about profit, market capital, revenue, and assets. Despite the strong push for innovation by the government, the top most priority of many enterprises is to build a stronger and larger enterprise, rather than creating real value for their clients. There seems to be something at play beyond the essential purpose of building a larger and stronger business empire.

Discussion and Conclusion

The Three Levels

The key themes emerging from the fieldwork are related to different levels of the cognitive frame, from the pattern of behavior, to the reasoning being used, to the deepest sources. Some of the themes may relate to the conscious mind, while others are more about subconscious aspects. By connecting the driving forces and the pattern of behavior, managers are able to see the structure – the story or the narrative – calling up an entire manifold tale of their cognition.

For my inquiry the most important thing is the recurring pattern, both verbal and non-verbal. Money and insecurity recur in verbal form, but they are also structured into reasoning and its foundations. Money is a recurring topic during the interviews. Managers tend to over-emphasize the importance of money, which is manifested in the behavior of changing jobs frequently and being enthusiastic about IPO. The narrative has to do with the purpose of life. It is taken for granted that during one's lifetime, the foremost things are those that can reveal value, such as money. There seems to be a driving force here – *homo economicus* – which gives shape to this behavior (see Table 3.1).

Table 3.1: The narrative – Example of money.

Levels		Example of money
Level 1	Event	1. Agitating for better pay. 2. In order to get better pay, employee is threatening to quit.
	A pattern	1. Working overtime for the entire week. 2. Constantly changing one's job for better pay and to show the value of one's life. 3. The fever for IPO.
Level 2	Reasoning (justification, structure)	1. Without money, everything is just empty words. 2. People can do nothing without money. 3. Everything relies on money. 4. The money earned through one's effort is good. 5. The salary earned proves the ability of a person. 6. With this money, the value of existence can be realized.
Level 3	Driving forces	1. The more I possess, the better life I will have. 2. Human nature: *homo economicus.*

Although these thoughts are manifested at different levels of a framework, the driving forces are holding them together. The reasoning is in the middle, connecting the driving forces with the behavior. For the convenience of communication and analysis we use these terms separately to describe the thought structure. Actually they form a single process.

The Fuzzy Logic System – Hindering or Enabling Transformation

The themes are operating under a system analogous to fuzzy logic. Driving forces recognize different scenarios and then respond in real time, ready to cope with different situations. The deepest sources prepare managers' reasoning for any situation they might face. The driving forces, along the themes, identify and handle the different situations that arise. Thus, the themes together form a fuzzy system, which operates automatically. This does not mean that in any situation all themes will be equally important. There will be different priority, different weights for them in different scenarios.

The majority of informants are not able to clarify this process, as this fuzzy logic system is operating at the subconscious level. There is a possibility that, by knowing about the mode of operation of this frame, most of their problem can be connected to its deepest sources. Then, they will be able to understand and appreciate the forces operating in the problematic ways they go about transformation.

The Sources of Managers' Cognition

From the findings, it can be seen that all the themes converge at a common foundation, although they are not completely congruent with each other at the reasoning and behavior levels. Looking more closely into the foundation, it would seem that managers select some concepts unconsciously and embed them in their framework: self-interest, Maslow's physical needs, and survival of the fittest. Further, materialism, individualism, and capitalism are terms that are currently popular in discourses of my informants, which become sources of inspiration for the managers in this study.

The sources of managers' cognition is guiding their managerial behaviors and individual life in such diverse aspects as the use of extrinsic incentives, strict internal control system, financial strategies by means of IPOs, buying more properties, or accumulating more wealth. These management practices and styles emanate from the managers' deepest conceptions regarding human nature, human relations, human activities, education, and organization itself. What is clear for me is that this foundation is creating a culture of competition and domination, with pattern of behavior characterized as plundering, speculating, or taking short cuts. Managers are less likely to care for the collective interest and, having less motivation for taking social responsibility, caring for the weak, or treating employees with sincerity. However, it is this fragile thread that binds the key themes together and shapes the ways they go about transformation.

Ultimate Sources: Quanzhou Managers

The concepts that are often relied upon, either consciously or subconsciously, by Quanzhou managers derive from various sources (see Table 3.2). They are most likely to be found in traditional culture and practices, the education system, Western value systems, the theory of scientific management, economic theories, military language, political movements, government policy such as the opening door and the reforms, and so on. Fleck (1981, p.46) argues that: "What actually thinks within a person is not the individual himself but his social community. The source of his thinking is not within himself but is to be found in his social environment and in the very social atmosphere he 'breathes.' His mind is structured, and necessarily so, under the influence of this ever-present social environment, and *he cannot think in any other way.*" Undoubtedly, all of these elements shape managers' pattern of behavior. Therefore, the policy of family firms in Quanzhou and its way of transformation are unquestionably shaped by such deepest sources as below.

Table 3.2: Ultimate sources.

Ultimate sources	Theme of transformation: Intellectual resource
Traditions	Family interest
Quanzhou local culture	Daring to struggle (*pinbo*, 拼才會赢)
The CCP's values	Economic development
Cultural Revolution and Political Movements	Competition, wolf culture, and the
May 4th Movement and Scientific Management	To be more dogmatic; a paradigm of reductionism
Economic Theories	Maximum welfare
Opening Door Era	The Cat Theory
Craze for Chinese Classics	No virtue education
Overseas Businessmen	The principle of *pinbo*
Learning By Doing	Copycatting, IPOs, CCTV advertisement
Business School Education	Legitimization of dehumanized style

Managers' motivations and behavior emanate from these deeply rooted sources, which they rarely think about and hence, much less likely to challenge them. Morgan (1986) describes the idea that 'organizations are ultimately created and sustained by conscious and unconscious processes, with the notion that people actually become imprisoned in or confined by the images, ideas, thoughts, and actions to which these processes give rise. The metaphor encourages us to understand that while organizations may be socially constructed realities, these constructions are often attributed an existence and power of their own that allow them to exercise a measure of control over their creators' (p. 207).

Over the last three decades of development of business management on mainland China, the elements listed in Table 3.2 have been the most important sources shaping the ways they go about work and transformation. It is like a traditional funeral in a Chinese village, Due to the lack of knowledge about life and death, people entrust the funeral arrangement to the master of rituals and let the master interpret and manage the situation as he wishes. Without questioning, they follow completely the procedures and rituals prescribed by the master. People are willing to accept this situation as nobody knows about death except for the master.

This is a moment when the villagers are placing their faith in those who know what to do. Why do they do that? They see that so many other people are showing their faith in the master, and hence they become believers as well. They are most comfortable in letting the masters run the show as they believe that they themselves do not possess certain knowledge which the masters have. In like manner, managers' cognition and their behavior tend to be guided by certain cherished attitudes and habits, widely accepted practices, theories, conceptions, and doctrines. Managers' fate seems to be in the hands of these masters, the orthodoxy. Managers will only be able to begin to understand their situation if they can take this further step to investigate the ultimate sources of their thoughts and behaviour.

The Continuum

To look into the thought structure means attention should be given to the deepest sources of their cognition. The visible problem is not the fault of managers, as they are just a part of the fuzzy system. Many managers tend to adopt, implement, and passively apply certain conceptions or principles to their daily management in the context of transformation. The deepest sources are manifested in the different styles of managerial behavior.

The changing thought structure is manifested in a continuum of certain themes, either staying as it is or being adjusted. That is why the attempt of this study is to describe this process, at a historical moment, as one possibility to look at. These themes are like a continuum or a thread with two ends, progressing from one to the other. Managers may choose to challenge the root causes of certain themes, make clear the deeply ingrained driving forces, enable the new knowledge to come in, and become more capable of discovering the root causes of the problem (see Table 3.3).

Table 3.3: The continuum.

A sound system	The continuum	A defective system
– Peace and joy	– A new way to diagnose the problem	– Use competing mindsets to solve the current problems
– Accompanying, sharing, collaborating	– Less competing, more collaborating	– Competing, manipulating
– Human is essentially spiritual	– Human needs to be more spiritual	– Human is essentially self-interested
– Sound foundation (spiritual capital; Zohar, 2004)	– Weakening the cherished conventions, by introducing new sources of knowledge such as spiritual capital	– Deficient foundation (*homo economicus*)

Different Levels of Managers' Cognition

In my setting there is considerable diversity. The informants have different backgrounds and various ages. It is not only one map, as different people appear to think differently and yet there is commonality that is very apparent. But, neither are the maps that emerge completely distinct from each other. There is an interplay of coherence and fragmentation that asks for a non-simplistic reading.

In Quanzhou, it is less likely for managers to have a coherent framework, even if an individual manager is looking for and hoping to have it. Maybe the thought structures in most subgroups are very similar. Maybe similarity resides at a deeper level, although there are minor differences on the surface. Maybe there are conflicts on the surface, but the driving forces in fact are congruent.

In Figure 3.1, small circles are used to identify the subgroups in this study's sample. Although there is similarity or overlap, they tend to have specific characteristics compared with each other. Subgroups are different in their patterns of behavior, but their driving forces tend to be the same, belonging to the same large circle. Thus, the small circles are all embedded in the larger one – the Quanzhou managers' framework.

Different subgroups may focus obsessively on specific themes. For example, owners stressed the importance of survival, resulting in a sense of insecurity. The second generation's thought structure seems to be more inclined towards taking risks. Although some informants are in more than one group, the source of his thinking is not within himself but is to be found in his social environment.

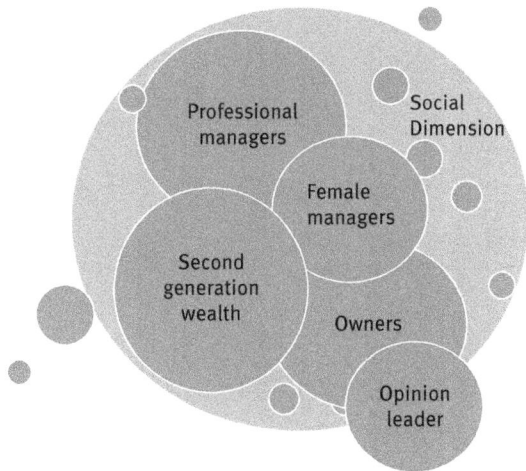

Figure 3.1: Different levels of Quanzhou managers' cognition.

Quanzhou Managers' Perceptions of Transformation

In China, in the past three decades, managing and transformation tends to be centered 'playing with capital', cash flow, *renmai* (human relations), struggling, and most recently, on the Internet. During this period, that perception about managing and transformation tends to vary over time and the latter is somehow fragmented, and its evolution is based on the experience and practice of entrepreneurs. In the transition, it is vital to look at these behaviors and structures, challenging the ontological thinking about management and organization and finally reach a deeper understanding about transformation (see Table 3.4).

Table 3.4: Quanzhou managers' perceptions of managing and transformation.

Themes	Management literature	Quanzhou managers' perception
Managing	– Command chain (Weber, 1947) – Value chain (Porte, 1985) – Scientific management (Fayol, 1949) – The human relations school (Mayo, 1933; Roethlisberger and Dickson, 1939) – Strategic planning (Ducker, 1994) – Contingency theory (Burns and Stalker, 1961; Lawrence and Lorsch, 1967) – Assumptions; assumptions of mission, environment and capacity, human nature (Drucker, 1994) – Strategic analysis (Porter, 1980) – Operations (Peters, 1993) – Forces and forms (Mintzberg, 1991) – A learning system (Senge, 2006) – Culture and leadership (Schein, 2004) – Spiritual capital (Zohar, 2004) – Execution (Welch, 2005) – Learning (Senge, 2006) – A thought system (Bohm, 1994)	– Making money – Internet – Family interest – Playing capital – *Renmai* (human relations) – Struggling – Battlefield

New Management Practices in the Context of Transition

Many of the local and global challenges facing Quanzhou managers today are embedded in interconnected systems of cognitive frame. Addressing these challenges means moving beyond the limitations of the perspectives, methods, and tools of traditional reductionist science. As Fritjof and Luisi (2014) put it, systems thinking is based on the fundamental shift of perception from the world as a dissociated

Table 3.5: New management practices of some exemplary companies.

Companies	Traditional practices	New practices
Xiaomi	KPI	De-KPI
	Bureaucratic	Flat
	High power distance	CEO is engineer
Fangtai	Competition	*Rushang* (Confucian businessman)
	Executive incentive schemes	*Shengguzhi* (a labor-sharing system for all employees)
Jingdong	Increase the movement of the whole society	Decrease the movement
	Profitability	Accept loss in order to improve the quality of service
Yunus	Client is not trustworthy	All clients should be trusted
	Profit shared by shareholders	Profit reinvested in the business
Wuyong	Cash flow management	No cash flow
	Marketing	No marketing
	Entrepreneurship	No title

collection of parts to the world as an integrated whole. Table 3.5 portrays the possible new management practices.

In summary, this study has analysed some selected managers' cognitive domain that broadly informs their work- and livelihood-related issues, aspirations, actions, and concerns in the context of ongoing reform and transition. The finding highlighted the different elements of managers' thought about transformation, including but not limited to the business practices of copycatting, initial public offering, playing capital, excessive reliance upon consultancy, and manufacturing upgrade with Internet plus. These then are the prevailing practices and ideas about innovation in Quanzhou managers' cognitive framework. Further, these elements have been shaped by its deepest sources, such as the Quanzhou local culture 'daring to struggle (*pinbo*, 拼才會贏)', the traditional concept about family, and the dominance of competition in the social environments. The entrepreneurs moulded by this cognitive framework tended to see their top most priority centered on immediate profit seeking, and transformation and innovation that do not lead to a clear and immediate profit making are not something entrepreneurs really like to do. However, if the needs or problems of the customer are not the first impulse for innovation, it is difficult to imagine that the outcome of their innovation would be accepted by the market in the long run. At the present, the entrepreneurs and managers in an under-sieged mentality are too pre-occupied with surviving in the fierce market. It is here that "perception is reality" is demonstrated in its full strength. Therefore, becoming conscious of and exploring the deepest sources within the managers' cognitive framework might be the first step for them to break away from their prevailing way of thinking, if they wish to do so at all. This study also serves to show the power of "self-fulfilling prophecy" and the difficulties in breaking away from the various elements inherent in their existing cognitive framework. Most importantly, it is hoped that this would help to supplement our understanding of the challenges and complexity that any attempts to bring about changes in management practices and thinking would have to face in places such as Quanzhou.

Appendix 1: Informant Pseudonyms and Background

	Pseudonym	Age	Gender	Position and company	Location of Interview
1	Kang Jing	45	Male	The CFO of a real estate company in Quanzhou	Restaurant
2	Shi Li	46	Female	The Human Resource Director of a villa developer in Quanzhou	Company
3	Liu Fang	48	Female	The CFO of a listed company, who is a senior financial affairs specialist	Company
4	Zheng Xin	42	Male	The Financial Controller of a real estate company	Company
5	Wang Xiang	45	Male	The Vice President of an auto sales company	Company and restaurant
6	Lu Hai	39	Male	The Vice President of a clothing company	Company
7	Chen Xing	72	Male	The founder and Chairman of a shipping company	Company
8	Dai Yi	45	Male	The owner of a family clothing business	Company and restaurant
9	Chen Liang	57	Male	The Vice President of a manufacturing company	Company and restaurant
10	Guo Yi	46	Male	The founder and CEO of a private company in the truck manufacturing industry	Company
11	Li Jin	37	Male	The CFO of a pre-IPO firm	Company and restaurant
12	Wang Qi	50	Male	One of owners of a tea business	Company

(continued)

	Pseudonym	Age	Gender	Position and company	Location of Interview
13	Liu Jun	40	Male	The Vice President of a manufacturing company	Company, teahouse, and restaurant
14	Wang Zong	55	Male	The owner of a consulting company	Teahouse, and restaurant
15	Chen Jin	47	Male	The owner and Managing Director of an incense business	Company
16	Pu Duan	45	Male	The owner of a clothing company	Company
17	Zhou Hua	40	Female	Assistant Chairman in a leather company	Company, Starbucks, and restaurant
18	Chen Qi	37	Male	The President of a shipping company	Company
19	Xie Jia	42	Male	The Vice President and one of the Founders of a wedding photo company	Company, Starbucks, and restaurant
20	Yang Yi	54	Male	The Chief Financial Official of FZ	Restaurant
21	Shao Sheng	37	Male	The Managing Director of a biomedical company	Starbucks
22	Chen Chang	52	Male	The owner of a consulting company	Company
23	Ding Hua	35	Male	The Vice President of a footwear firm	Company, Starbucks, and restaurant
24	Yang Feng	37	Male	The Chief Financial Official of a leather enterprise	Company
25	Huang Zi	52	Male	The General Manager of a listed transportation company	Company
26	Lai Chang	36	Male	The Finance Manager of a listed company	Voice chat in WeChat
27	Lin Jin	37	Male	The Human Resource Manager of a real estate company	Voice chat in WeChat and restaurant
28	Liu You	37	Female	The Accounting Manager of a sportswear firm	Voice chat in WeChat
29	Wu Mei	32	Female	The Vice President of a listed leather enterprise	Company
30	Liu Zhen	45	Male	The Vice President of an electronic enterprise	Restaurant
31	Li Xing	53	Male	The CFO of a listed leather company	Company and restaurant
32	Liu Dong	48	Male	The Director of Procurement of a listed leather company	Company

(continued)

	Pseudonym	Age	Gender	Position and company	Location of Interview
33	Shi Jin	37	Male	The General Manager of Investment and Internal Control of a listed leather company	Teahouse, and restaurant
34	Xiao Pei	44	Male	The owner of a trading firm	Restaurant
35	Ye Li	43	Female	The owner of an import and export company	Restaurant
36	Yan Shu	32	Female	The Section Head in the front office of a five star business hotel	Restaurant
37	Lu Ying	57	Male	The founder and Chairman of a biomedical enterprise	Company and restaurant
38	Zhang Guo	45	Male	A construction project manager	Restaurant
39	Ding Ying	53	Female	The Board member of a biomedical enterprise	Company and restaurant
40	Jiao Ying	53	Male	The Board member of a biomedical enterprise	Company and restaurant
41	Lin Shui	56	Male	The founder of a footwear manufacturing company	Company
42	Wu Ji	33	Male	The Financial Controller of a footwear manufacturing company	Company
43	Cui Lin	44	Female	The owner of a furniture manufacturing enterprise	Starbucks
44	Ye Xiang	52	Male	The Vice President of a pre-IPO company	Company and restaurant
45	Huang Zhong	56	Male	The owner of a garment factory	Restaurant and teahouse
46	Yang Hong	40	Male	The Vice President of a listed transportation company	Company and restaurant
47	He Sha	55	Male	The founder of a private business	Restaurant
48	Liu Hong	45	Male	An administration official	Company and restaurant
49	Ke Hui	46	Male	The General Manager of a Petro Trading company	Company
50	Lin Rong	60	Female	The founder of a landscaping firm	Company

References

Agar, M. (1996). *The Professional Stranger: An Informal Introduction to Ethnography*, San Diego: Academic Press.

Argyris, C. (1976). *Increasing Leadership Effectiveness*. New York: John Wiley.

Argyris, C. (1982). *Reasoning, Learning, and Action: Individual and Organizational*. San Francisco, CA: Jossey-Bass.

Bartunek, J. M. (1993). The multiple cognitions and conflicts associated with second order change. In J. K. Murnighan (Ed.), *Social Psychology in Organizations: Advances in Theory and Research*, pp. 322–349. Englewood Cliffs, NJ: Prentice Hall.

Bohm, D. (1994). *Thought as a System*. London and New York: Routledge.

Burns, T. & Stalker, G.M. (1961). *The Management of Innovation*. Tavistock Publications, London.

Caijinglangyan (2016), *Hulianwang Meiyou Dianfu Shenme* (In Chinese). Retrieved from http://video.tudou.com/v/XMTcxNDM3MjEwOA==.html.

Capra, F. & Luisi, P. L. (2014). *The Systems View of Life: A Unifying Vision*. Cambridge: Cambridge University Press.

CNBC (2016), China's economy grew 6.9 percent in 2015, a 25-year low. Retrieved from https://www.cnbc.com/2016/01/18/china-reveals-key-q4-2015-gdp-data.html.

Douglas, M. (1986). *How Institution Think*. New York: Syracuse University Press.

Drucker, P. F. (1994). The Theory of the Business, *Harvard Business Review*, September-October, 95–104.

Einstein, A. (1946). Atomic Education Urged by Einstein, *New York Times*, May 25, p.13.

Faure, G.O. & Fang, T. (2008). Changing Chinese values: Keeping up with paradoxes. *International Business Review*, 17, 194–207.

Fayol, H. (1949). *General and Industrial Management*. Translated by C. Storrs, Sir Isaac Pitman and Sons, London.

Fleck, L. (1981). *Genesis and Development of a Scientific Fact*. Chicago: University of Chicago Press.

Forrester, J. W. (1971). Counterintuitive behavior of social systems. *Technology Review*, 73(3), 52–68.

Garvin, D., & Roberto, M. (2005). Change through persuasion. *Harvard Business Review*, 83(2), 105–112.

Gavetti, G., & Levinthal, D. (2000). Looking forward and looking backward: Cognitive and experiential search. *Administrative Science Quarterly*, 45(1), 113–137.

Gilbert, C. G. (2006). Change in the presence of residual fit: Can competing frames coexist? *Organization Science*, 17(1), 150–167.

Gioia, D. A., & Chittipeddi, K. (1991). Sensemaking and sensegiving in strategic change initiation. *Strategic Management Journal*, 12, 433–448.

Goffman, E. (1974). *Frame Analyses: An Essay on the Organization of Experience*. Cambridge, MA: Harvard University Press.

Harvey, D. (2005). *A Brief History of Neoliberalism*, New York: Oxford University Press.

Huaxiajishi (2016), *Caogen Nixi de Ding Shizhong, Yibaiyi Caiganggang Kaishi* (In Chinese). Retrieved from http://news.efu.com.cn/newsview-1149983-1.html (2012).

Huff, A. S. (1990). *Mapping Strategic Thought*. Chichester: John Wiley and Sons.

Isaacs, W. (1999), *Dialogue*, New York: Currency.

Cornelissen, J., & Werner, M. (2014). Putting framing in perspective: A review of framing and frame analysis across the management and organizational literature. *Academy of Management Annals*, 8(1), 181–235.

Kaplan, S. (2008). Framing contests: Strategy making under uncertainty. *Organization Science*, 19(5), 729–752.

Kotter, J. (1996). *Leading Change*. Boston, MA: Harvard Business School Press.

Lawrence, P.R. & Lorsch, J.W. (1967). Differentiation and integration in complex organizations. *Administrative Science Quarterly*, 3(1), 1–47.

Lewin, K. (1947). *Field Theory in Social Science*. New York: Harpers & Row.

Ma, Guangyuan (2015), *Caijinglangyan: Zhongguo Zhizao 2025* (In Chinese). Retrieved from http://www.iqiyi.com/v_19rrolrxl0.html.

Madsen, Richard. (2011). Religious renaissance in China today. *Journal of Current Chinese Affairs*, 40(2), 17–42.

Mantere, S., Schildt, H., & Sillince, J. (2012). Reversal of strategic change. *Academy of Management Journal*, 55(1), 172–196.

Mayo, E. (1933). *The Human Problems of an Industrial Civilization*. Cambridge, MA: Harvard Business School.

McGregor, D. M. (1960/1985). *The Human Side of Enterprise*, New York, NY: McGraw-Hill.

Metcalfe, J., & Shimamura, A. P. (Eds.). (1994), *Metacognition: Knowing about Knowing*. Cambridge, Mass: The MIT Press.

Meyer, R. E., Hollerer, M. A., Jancsary, D., & Van Leeuwen, T. (2013). The visual dimension in organizing, organization, and organization research: Core ideas, current developments, and promising avenues. *Academy of Management Annals*, 7(1): 489–555.

Mintzberg, H. (1989). *Mintzberg on Management: Inside Our Strange World of Organizations*. New York: Free Press.

Mintzberg, H. (1991). The effective organization: Forces and forms. *Sloan Management Review*, 32(2), 54–67.

Morgan, G. (1986). *Images of Organization*. Newbury Park, CA: Sage.

Nadkarni, S., & Narayanan, V. K. (2007). Evolution of collective strategy frames in high and low velocity industries. *Organization Science*, 18(4), 688–710.

Osburg, J. (2013), *Anxious Wealth: Money and Morality among China's New Rich*. Stanford: Stanford University Press.

Peters, T. (1993). *The Tom Peters Seminar: Crazy Times Call for Crazy Organizations*. London: Macmillan.

Porter, M.E. (1980), *Competitive Strategy*. New York: Free Press.

Porter, M. E. (1985). *Competitive Advantage: Creating and Sustaining Superior Performance*. New York: Simon and Schuster.

Redding, S. G. (1993). *The Spirit of Chinese Capitalism*. Berlin: de Gruyter.

Reed, G. G. (1991), *The Lei Feng phenomenon in the People's Republic of China*, Unpublished doctoral dissertation, University of Virginia.

Roethlisberger, F.J., & Dickson, D.W. (1939). *Management and the Worker*. Cambridge, MA: Harvard University Press.

Schein, E.H. (2004). *Organizational Culture and Leadership (3rd Ed.)*, San Francisco: Jossey-Bass.

Senge, P. (2006). *The Fifth Discipline*. New York: Doubleday.

Tang, J. & Ward, A. (2003). *The Changing Face of Chinese Management*. New York: Routledge.

Tisdell, C. (2009). Economic reform and openness in China: China's development policies in the last 30 years. *Economic Analysis and Policy*, 39(2), 271–294.

Vernezze, P. (2011). *Socrates in Sichuan, Chinese Students Search for Truth, Justice, and the (Chinese) Way*. Washington, D.C.: Potomac Books.

Wang, H. (2003). *China's New Order: Society, Politics, and Economy in Transition*. Translated by T. Huters and R. E. Karl. Cambridge, Mass: Harvard University Press.

Warner, M. (2014). *Understanding Management in China*. London and New York: Routledge.

Weber, M. (1947). *The Theory of Social and Economic Organization*. Translated by A.M. Henderson and T. Parsons. New York: Oxford University Press.

Weick, K.E. (1995). *Sensemaking in Organizations*, Thousand Oaks, CA: Sage.

Welch, J. & Welch, S. (2005), *Winning*. New York: HarperCollins.

Wheatley, M. J. (2006), *Leadership and the New Science: Discovering Order in a Chaotic World, (3rd ed.)*. San Francisco: Berrett-Koehler.

Witzel, M. (2012), *A History of Management Thought*, New York: Routledge.

Zhang, D.S. (1997), *Si Ru Feng Yun: Xian Dai Zhongguo de Si Xiang fa Zhan Yu She Hui Bian Qian* (In Chinese). Taipei: Ju Liu Tu Shu Gong Si.

Zhang, Lifan (2016), *Bixu Jieshou Pushijiazhiguan, Foze Nijiubushuyu Quanrenlei* (In Chinese). Retrieved from http://www.360doc.com/content/16/0107/06/19446_526057941.shtml.

Zhao, W. & Arvanitis, R. (2010). The innovation and learning capabilities of Chinese firms. *Chinese Sociology and Anthropology*, 42(3), 6–27.

Zohar, D. (2004). *Spiritual Capital*. San Francisco: Berrett-Koehler.

Part III: **Entrepreneurship, Learning and Innovation**

Xi Zhao, Jacky Hong and Robin Snell

4 Entrepreneurial Ecosystems in China

Introduction

The entrepreneurial ecosystem catalyzes innovations and regional economic growth, thus, scholars and practitioners deliberately cultivate it (Adner & Kapoor, 2010; Mason & Brown, 2014; Spigel & Harrison, 2017). The entrepreneurial ecosystem involves dynamic interaction among various entrepreneurial actors, such as entrepreneurial firms, established companies, professional investors, social institutions, and governments, to promote innovation and revitalize the economy (Spigel & Harrison, 2017). The entrepreneurial ecosystem perspective complements regional development literature and strategy literature, moreover, entrepreneurship serves as an input, as well as an outcome of the entrepreneurial ecosystem Therefore, entrepreneurial ecosystem presents co-evolutionary patterns (Acs et al., 2017). By confronting liabilities of smallness and newness, entrepreneurial firms can gain access to local resources, legitimacy, reduced risks, and accelerate innovation in the entrepreneurial ecosystem (Adner & Kapoor, 2010; Eisenhardt & Schoonhoven, 1996; Spigel & Harrison, 2017). There is vibrant research on the formation and performance of the entrepreneurial ecosystem.

Prior literature examines innovative outcomes to evaluate the performance of entrepreneurial ecosystems (Adner & Kapoor, 2010; Dougherty & Dunne, 2011). However, dynamic and self-evolvement of the entrepreneurial ecosystem result in difficulties to measure the effectiveness (Isenberg, 2011; Mason & Brown, 2014; Spigel, 2015; Spigel & Harrison, 2017). This paper will adopt the attribute framework of the entrepreneurial ecosystem to illustrate the entrepreneurial ecosystem in China and identify how entrepreneurial ecosystems affect the innovation process.

This paper contributes to the entrepreneurial ecosystem literature in two ways. First, regional development literature and strategy literature ignores the role of entrepreneurial firms, thus, this paper adopts an entrepreneurial ecosystem perspective to identifies cultural, social, and material attributes of entrepreneurial ecosystems in China. It has shown top three entrepreneurial ecosystems in China are driven by different attributes to facilitate their evolvement. Second, this paper draws from the entrepreneurial ecosystem perspective to explain how entrepreneurial ecosystems affect the innovation process. It identifies and discusses the various configuration of an entrepreneurial ecosystem affect invention and commercialization activities.

The rest of this paper consists of four sections. The first section will present the background of the entrepreneurial ecosystem. In providing a framework of entrepreneurial ecosystems, the second section will discuss attributes of entrepreneurial ecosystems in China. Zhongguancun (ZGC), Zhangjiang (ZJ), and Nanshan (NS) are the top three innovative entrepreneurial ecosystems in China and maintain

https://doi.org/10.1515/9783110715002-004

various advantages to promote innovation. The third section will provide an illustration of the innovation process in entrepreneurial ecosystems. The last section will summarize contributions and point out implications.

Background of Entrepreneurial Ecosystem

The ecosystem is a biological term and describes how living organisms interact in a complex context (Tansley, 1935). Successful entrepreneurship does not occur in a vacuum, and it shapes and is shaped by the wider community context (Mason & Brown, 2014). Social, cultural, and institutional elements collectively affect the emergence of the entrepreneurial ecosystem. Scholars introduced the concept of the ecosystem into the business world to emphasize the localized context to produce interdependent relationships among business entities. "Business ecosystems condense out of the original swirl of capital, customer interest, and talent generated by a new innovation, just as successful species spring from the natural resources of sunlight, water, and soil nutrients" (Moore, 1993, p.76). The ecosystem is the first used in the academic literature in the study of Silicon Valley to describe the closely connected relationship among universities, research institutions, companies, professional institutions, and government "feed off, support, and interact with each other" (Bahrami & Evans, 1995, p.63). Overall, the entrepreneurial ecosystem involves multiple entrepreneurial actors who share common goals and co-evolve overtime.

Scholars define an entrepreneurial ecosystem consist of various actors and functions, and co-located elements interdependent with each other to facilitate the growth of new ventures and drive the prosperity of the ecosystem (Adner & Kapoor, 2010; Isenberg, 2010; Stam, 2015; Spigel & Harrison, 2017). However, there is no commonly accepted definition of the entrepreneurial ecosystem (Malecki, 2017). The entrepreneurial ecosystem is dynamic and self-evolving with complex interactions among actors. Mason & Brown (2014, p.5) provide a comprehensive definition of the entrepreneurial ecosystem as a "set of interconnected entrepreneurial actors, entrepreneurial organizations, institutions and entrepreneurial process which formally and informally coalesce to connect, mediate and govern the performance within the local entrepreneurial environment".

The entrepreneurial ecosystem is an umbrella concept, and scholars identify two themes in entrepreneurial ecosystem literature (see Table 4.1). The first theme focuses on the entrepreneurial environment, in other words, creating a fully functional environment drives the success of innovation and entrepreneurship. Isenberg (2011) identified six domains of the entrepreneurial ecosystem to promote entrepreneurial performance. World Economic Forum (2013) presents eight pillars of the entrepreneurial ecosystem to facilitate the integration of resources.

Table 4.1: The constitution of entrepreneurial ecosystems.

Theme	Elements	Authors
Environments promotes or restrict the entrepreneurship	Six domains: Policy, culture, supports, human capital and market, of the entrepreneurial ecosystem	Isenberg, 2011
	Eight pillars: Accessible markets, human capital, funding & finance, support systems, government & regulatory framework, education & training, major universities as catalysts and cultural support	World Economic Forum, 2013
The interdependency relationship drives the entrepreneurship	Seven components: Informal network; formal network; university; government; professional and support services; capital services; talent pool.	Cohen, 2006
	Embryonic ecosystems and scale-up ecosystem	Mason & Brown, 2014
	Eleven elements: Networks; leadership; finance; talent; knowledge; support services/intermediaries; demand; physical infrastructure; culture; formal institutions.	Stam, 2015
	Nine attributes: Leadership, intermediaries, network density, government, talent, support service engagement, companies and capitals.	Feld, 2012
	Three categories of attributes: cultural, social and material	Spigel, 2015

The second theme recognizes the importance of entrepreneurial agency, and scholars switch the attention to the relationship and interaction between actors and the environment. The entrepreneurial ecosystem integrates and promotes knowledge exchange and recycle (Spigel & Harrison, 2017). The knowledge flow supports the formation of an interdependency relationship in the entrepreneurial ecosystem and promotes the self-evolving of the entrepreneurial ecosystem.

Cohen (2006) identified seven components to contribute to a sustainable entrepreneurial ecosystem. Mason & Brown (2014) classified entrepreneurial ecosystems into embryonic ecosystems and scale-up ecosystems. Comparing with the scale-up ecosystems, embryonic ecosystems consist of startups with limited interaction ecosystems. Thus, startups perceive limited benefit from the construction and dissemination of knowledge from business entities in the ecosystem. Stam (2015) focused on the causal relationship of identified elements. For example, formal institutions, culture, exogenous demand, and physical infrastructure are the fundamental causes of value creation. Networks of entrepreneurs, leadership, finance, talent, knowledge, and support service are the fundamental causes of entrepreneurial activities and determine the success of the entrepreneurial ecosystem. However, identified elements largely overlap and scholars cannot reach an agreement on the mixture of elements in the entrepreneurial ecosystem (Theodoraki et al., 2018). Feld (2012) and Spigel (2015) examine attributes of the entrepreneurial ecosystem, and Spigel (2015) categorizes ten attributes into three types, social, culture, and material.

Embedded in entrepreneurial ecosystem associates various resources which promote and impede the innovation process in new ventures, corporate ventures, and established companies (Welter, 2011; Zahra & Nambisan, 2012). "The advantages of an entrepreneurial ecosystem are related to resources specific to the entrepreneurship process" (Spigel, 2015, p.4). Drawing on three types of attributes, the paper will follow Spigel (2015) in adopting three types of attributes, that is, cultural, social, and material, to discuss the landscape of the entrepreneurial ecosystem in China and how do entrepreneurial ecosystems promote innovation (see Figure 4.1).

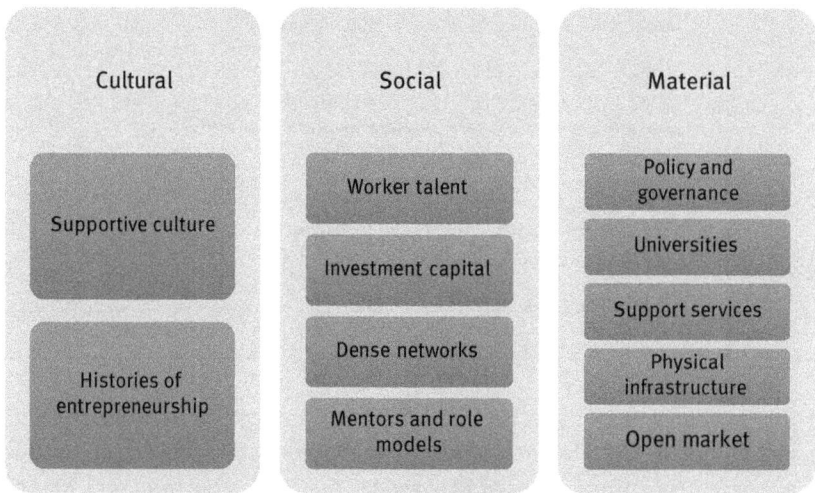

Figure 4.1: Framework of entrepreneurial ecosystem attributes.

Cultural attributes refer to a shared culture that encourages entrepreneurial actors to contribute efforts (Spigel, 2015). The entrepreneurial culture facilitates entrepreneurial actors to overcome bounded rationality, present role models and encourage entrepreneurial actors to take risks and create value (Cohen et al., 2018; Spigel, 2015; Spigel & Harrison, 2017). The entrepreneurial culture is the premise for entrepreneurs to actualize entrepreneurial opportunities.

Except for cultural attributes, the entrepreneurial ecosystem enables entrepreneurial actors to access various resources, including worker talent, investment capital, networks, and mentors (Spigel, 2015). The entrepreneurial ecosystem supports the circulation of technologies, skills, talent, market knowledge, and financial capital. The strong connections in the entrepreneurial ecosystem facilitate knowledge assimilation and commercialization (Adner & Kapoor, 2010). Mentors and role models facilitate entrepreneurial knowledge cycling and enhance performance of the ecosystem (Spigel, 2015; Spigel & Harrison, 2017).

In addition, material attributes are physical subjects, including policy and governance, universities, support services, physical infrastructure and open market (Spigel, 2015). The presence of material attributes provides specialized support to regulate operation mechanisms within the entrepreneurial ecosystem. Thus, material attributes present formalized policies and markets to coordinate the collaboration and competition in the entrepreneurial ecosystem (Kapoor & Lee, 2012).

The Landscape of Entrepreneurial Ecosystem in China

The formation of an entrepreneurial ecosystem promotes knowledge exchange and innovation exploration. Various geographic and socio-cultural elements shape the configuration of the entrepreneurial ecosystem. Thus, this paper will focus on the most three innovative entrepreneurial ecosystems in China (Xinhua, 2016), that is Zhongguancun (ZGC), Zhangjiang (ZJ), and Nanshan (NS), to discuss the attributes of entrepreneurial ecosystems in China (see Table 4.2).

Table 4.2: Attributes of entrepreneurial ecosystems in China.

	ZGC	ZJ	NS
Cultural attributes	Cultural orientation for breakthrough innovation	Cultural orientation toward learning	Culture support for entrepreneurship
Social attributes	Networks oriented towards local academic institutions, social institutions, and SOEs	Networks oriented towards joint-venture corporations and develop innovative capabilities	Aggregating innovative resources and scaling bricolage for startups
Material attributes	Prestige research institutions attract talents and develop the technology. Governance policies promote institutional change and upgrade entrepreneurial services. A strong market promotes the actualization of entrepreneurial opportunities.	Specific policies facilitate the formation of physical infrastructure. Industrial chain formation provides networking and collaborating networks.	A strong market creates entrepreneurial opportunities. Market-oriented policies facilitate governance mechanisms and improve entrepreneurial services and infrastructure.

Zhongguancun (ZGC), Beijing

ZGC located in the Haidian district of Beijing and it is the first national high-tech industrial development zone in China. The State Council put forward the fundamental policy of socialist modernization construction and emphasis the scientific and technological development should be oriented to economic construction. During the widespread economic reform, Chenchunxian established the first private company in China and promote the formation of ZGC in 1980 (Chen, 2003). The emergence of "Zhongguancun Electronics Street" is the hallmark of the economic construction. The State Council established Beijing New Technology Industrial Development Trial Zone in 1988, and ZGC became the first high-tech park in China. The State Council launched a series policy and plans to support the development of ZGC (China Daily, 2020; ZGC Profile, 2021) (see Table 4.3).

Table 4.3: Milestones of ZGC.

	Milestones
May, 1988	Beijing New Technology Industrial Development Trial Zone
June, 1999	Provisional Regulations of Beijing Municipal New Technology Industry Development Pilot Zone
March, 2009	the construction of the Zhongguancun National Demonstration Zone
Jan, 2011	Development Plan Outline for Zhongguancun National Demonstration Zone

Moreover, ZGC has the most talented individual and scientific resource in China, and it is called China's Silicon Valley (ZGC Profile, 2021). Prestigious universities and research institutions are key actors to produce technologies and foster talented (Spigel, 2015). Tsinghua University, Peking University, and other 40 universities enable the stable supply of the most talented individuals. Meanwhile, more than 200 national research institutions, including the Chinese Academy of Science and the Chinese Academy of Engineering, stimulate technology transfer (Li, 2019). Over 200 business incubators and professional institutions provide connections, space, workshops, and investment for Chinese startups (China Daily, 2020; Li, 2019). The dense academic atmosphere promotes fundamental research and independent innovation, meanwhile, Beijing provides comprehensive and explicit intellectual property protections for business entities (Li, 2019). ZGC focuses on building State Key Laboratories and promote technological development, including the Three Gorges Project, and the Qinghai-Tibet Highway construction (China Daily, 2020). The well-developed physical infrastructure and institutional connections allow knowledge to spill over within the entrepreneurial ecosystem.

Social, corporate, and private institutions improve entrepreneurial services and coordinate entrepreneurial resources. 250 incubation institutions and 670 venture capital institutions located in ZGC to accelerate the coordination of entrepreneurial resources and knowledge flow (Gu & Sha, 2021). ZGC advocates collaboration across the state-owned enterprises (SOEs), universities, and research institutions to promote commercialization activities (Fan, 2018; Zhao & Pira, 2013). For example, Tsinghua Unigroup is the subsidiary of Tsinghua University and its Enterprise Resource Planning system enhances the firm's competence. Meanwhile, Tsinghua Unigroup benefits from the commercialization activities (Zhuang, 2015). Breakthrough innovations, including supercomputer, and aerospace engineering, take placed in ZGC. ZGC accounts for one-third of the investment amount of venture capital (China Daily, 2020). Meanwhile, enterprises in ZGC participate in international standards, national standards, and industrial standards formulation. Technology transaction values exceed one-third of the state (China Daily, 2020).

Over 30 years of development, ZGC has attracted entrepreneurial firms in the high-tech sector and contributes 25% of GDP in Beijing (Fan, 2018). More than 20,000 technology companies, such as JingDong, Lenovo, Baidu, Xiaomi, and iQiyi, located in ZGC. There are 71 companies publicly traded in the overseas market (Gu & Sha, 2021). With a successful career in entrepreneurship, ZGC is the hometown of Chinese unicorns like ByteDance, Didi, and Meituan (ZGC Index, 2021). TikTok and Douyin are the best products of ByteDance and the company worth over $100 billion in eight years (CCIDNET, 2020). The successful entrepreneurial stories shape the perceptions of existing and potential participants and drive the evolvement of the entrepreneurial ecosystem. According to Fan (2018), all informants in Beijing believe Beijing will be the engine of the high-tech sector and joint laboratories remove barriers to commercialize scientific research with companies.

Overall, ZGC presents a strong ingenious innovative capability to transfer knowledge from research institutions and promote commercialization activities with a local knowledge base. The advanced ZGC entrepreneurship ecosystem was established with the help of elite educational institutions, effective resource coordination, and commercialization.

Zhangjiang High-Tech Park (ZJ), Shanghai

Zhangjiang High-Tech Park (ZJ) was established in 1992 in Shanghai. The preferential geographic advantage of ZJ attracts overseas R&D investment, financial capital, and talents. Shanghai positions as the global innovation and technology hub and ZJ actualizes the plan by releasing policies and fostering an inclusive culture to upgrade the entrepreneurial ecosystem. It has 18,000 registered companies and 53 headquarters of multinational companies (Zhangjiang Profile, 2021).

ZJ is dominated by electronic, technology, and the biomedicine sectors, and aims to build a world-class integrated circuit (IC) cluster. The clear position enables ZJ to concentrate resources to develop the industrial chain to support the IC-focused entrepreneurial ecosystem. For example, SMIC is the leading IC it foundry enterprise and established headquartered in ZJ (SMIC, 2021). Huahong Group performs IC manufacturing business and facilitates industrial chain development (Huahong Group, 2019).

In the first phase of the development of ZJ, the Shanghai Municipal Government developed the "Focusing on Zhuangjiang" policy to provide preferential treatment and flexible regulation to aggregate local resources. Thus, Shanghai is the first city for foreign companies to tap into the Chinese market, thus, joint ventures land in ZJ (China Daily, 2019). 484 foreign companies establish headquarters in Shanghai (Tang, 2015). Six of the top 10 global chip designers, Qualcomm, Broadcom, NVIDIA, AMD, Marvell in ZJ (Su, 2019). Meanwhile, Shanghai Huahong NEC built the first 8-inch production line in 1999, and SMIC built the first 12-inch production line in 2005. Meanwhile, Huali Microelectronic operates the first 30mm fully automated wafer fab in mainland China and provides services for global clients (HLMC, 2020). Huali adopted the Original Equipment Manufacturing (OEM) business model to produce and assemble electronic devices. ZJ IC industrial cluster is formed.

Since 2004, ZJ established IC industrial chain including chip design, manufacturing, wafering testing, and packaging and has made breakthroughs in many fields such as mobile communication and digital TV. For example, Comlent Technology has developed China's first mobile phone RFIC transceiver chip (Microwave radio network, 2004), Telegent Systems produced 100 million mobile TV receiver chips (Telegent Systems, 2010). Foreign companies cooperate with local institutions to establish business incubators. For example, Johnson & Johnson launched an incubator for healthcare startups (Xinhua, 2019a). The regional headquarter of Plug and Play lands in ZJ (Xinhua, 2019b). Microsoft AI & IoT Insider Lab was launched in 2019 and aims to train engineers to provide cloud service and AI solutions (He, 2019). Global incubators bring innovative ideas to boost technological innovation for ZJ. ZJ forms alliances to encourage cooperation between entrepreneurial firms and large companies and fosters new projects to transform innovative projects from large companies (China Daily, 2019).

The growth phase starts in 2009. China has become an important IC manufacturer and consumer country in the Asia-Pacific region by virtue of its large market demand, low production cost, and abundant human resources and policy support. Moreover, ZJ works well in attracting foreign investment and receives $22.86 million in foreign investment in 2019 (Xinhua, 2019b). Meanwhile, ZJ creates 60 percent of foreign trade value (China Daily, 2019). The involvement of the municipal government set up government-supported venture funds and encourage technology transfer. ZJ takes the initial actions to transfer the "real estate manager" to a technological venture investor in 2014. In other words, ZJ serves as angel investors to provide initial funding as well as property, preferential policies, and comprehensive services (Wang, 2017). Corporations in ZJ closely keep up with the most advanced technology through partnership and promote scientific and technological innovation (Tang, 2015). Corporations develop indigenous innovative capability, the comprehensive elements of the industrial chain.

Overall, the geographic advantage and preferential policies play a critical role in importing, transferring, and creating knowledge in ZJ. Creating differentiated value propositions for foreign partners, companies in ZJ actualize opportunities through cooperative partnerships.

Nanshan Technology Park (NS), Shenzhen

Nanshan Technology Park (NS) was established in 2001 in Shenzhen, and it is market-oriented to promote high-tech companies (Shenzhen Nanshan, 2013). There are 171 public companies, which account for 40 percent of total companies in NS (Liu, 2021).

Shenzhen has a strong sense of reform and innovation and is the first Economic Development District. In the 1980s, China was transformed from a planned economy to a reform and opening-up. With the huge flow of migrants, 75 percent of permanent residents are immigrants, thus, Shenzhen is the city of immigrants (Shenzhen Daily, 2020). According to a survey conducted by the Chinese University, Hong Kong Baptist University, and some other Hong Kong universities, 16% of adults in Shenzhen were entrepreneurs in 2016 (Yang, 2018). Shenzhen's culture respects independent and innovative entrepreneurial efforts. Large corporations in Shenzhen encourage innovation and have a high tolerance for failure. Moreover, startups can easily receive capital from corporate venture capital, such as Tencent (Jin, 2020; Yang, 2018).

Shenzhen promotes an innovative culture. It opens the first industrial zone of Shekou, establishes the first commercial bank, China merchant bank, launches the first insurance company, Pingan Insurance. Shenzhen actively leverages the advantage of the Economic Development District and creates a well-development infrastructure to promote human, capital, and technology flow. In order to promote innovative resources, NS creates a better living and development space for talents and corporations. Shenzhen takes initiative to establish the first virtual university to attract talents from prestigious universities and research institutions in 1999. The virtual university consists of 53 universities and 42 research institutions in China and overseas.

Guided by servant leadership, the Shenzhen government provides benefits for high-tech corporations to promote a vigorous entrepreneurial ecosystem (Dang & Yan, 2020). Shenzhen government issues favorable incentive policies for startups, in regard to loans and taxes. For example, most startups are in the IT industry and they have light assets to leverage financial capital to expand the business. The government investment relaxes financial requirements and encourages the growth of high-potential corporations. NS has 33 cooperate financial institutions and 8 of them focus on startup financing. Local financial institutions promote specified loans for corporations at incubation, growth, and pre-New Three Board (Ma & Zeng, 2017). Government and other public motivate entrepreneurial passion by simplifying the administrative process. For example, it grants green passes for public companies to accelerate the process of uploading documents. Meanwhile, it adopts a new process to simplify the file filling process. Besides that, Shenzhen releases innovative policies to address livelihood issues. According to a report on the quality ecosystem for talent innovation and entrepreneurship, Nanshan ranks first among first-tier cities in terms of talent living and development environment (Lei, 2018).

NS provides a series of subsidies policies, government investment, and financial leverage to aggregate complementary resources for entrepreneurial actors to acquire knowledge, resources, and markets. Large leading corporations, such as Huawei, Tencent, and DJI, provide opportunities and resources to promote commercialization activities. Shenzhen, numerous hardware factories clustered and produce electronic parts for leading companies. Tightly coupled entrepreneurial resources provide advantages and convenience for startups to leverage complementary resources from leading corporations to conduct commercialization activities. NS has six unique features, (1) 90 percent of innovative companies are local companies; (2) 90 percent of research institutions are located in companies; (3) 90 percent of researchers work in companies; (4) 90 percent of research funding comes from companies; (5) 90 percent of intellectual property right belongs to companies; (6) 90 percent of high-tech intellectual property rights and projects are undertaken by companies, to distinguish from other ecosystems (Sun & Jing, 2021). The Shenzhen-Hong Kong stock connect program facilitates the capital flows from Hong Kong and accelerates the commercialization activities by integrating regional hardware manufacturing capabilities (Jing, 2019).

Overall, three entrepreneurial ecosystems in China present various features and configurations. Academic knowledge is the fundamental element to facilitate independent innovation in ZGC. Geographic advantages of ZJ result in attractiveness for global leading corporations. ZJ built a comprehensive industrial chain to promote advanced knowledge transfer and commercialize innovative products. NS forms entrepreneurial culture to aggregate leading companies to feed the entrepreneurial ecosystem.

Impact on Innovation Process

The emergence and development of an entrepreneurial ecosystem drive the resource flow and entrepreneurial opportunities (Spigel, 2015). Thus, operating in the entrepreneurial ecosystem shapes entrepreneurial behaviors and affects the innovation process. The process of innovation comprises invention and commercialization (Adner, 2006; Schumpeter, 1934). Given the uniqueness of entrepreneurial ecosystems in China, the following section will discuss how the entrepreneurial ecosystem affects the different forms of innovation.

Zhongguancun (ZGC), Beijing

ZGC has well-functioning material attributes to maintain effective operation in the entrepreneurial ecosystem, meanwhile, material attributes facilitates the invention and commercialization of innovation process (See Figure 4.2). Firstly, it aggregates prestigious universities and research institutions to conduct fundamental research and generate novel ideas. Meanwhile, "research engines" associate with intellectual properties, and collaborate with social institutions to commercialize market opportunities based on explicit intellectual property protections (Fan, 2018). Prestige universities and research institutions are keystone players to maintain the innovation architecture and generate innovative products and services. The R&D and knowledge spillover drive opportunities for commercialization in the entrepreneurial ecosystem. Secondly, the emergence of supportive institutions serves as an orchestrator to facilitate knowledge flow and coordinates resources within the entrepreneurial ecosystem. The government release series of industrial and technological policies to promote indigenous innovation.

The well-functioning material structure is regarded as the fundamental cause of social and cultural attribute formation. Entrepreneurial resources are widely available in the entrepreneurial ecosystem (Spigel & Harrison, 2017; Zahra & Nambisan, 2012). Entrepreneurial actors and resources, such as government, incubators, professional services, and investment, promote resource scanning and products diffusing. Meanwhile, startups, corporate ventures, and established corporations gathered in ZGC and build competencies to upgrade existing skills as well as commercialize emerging technologies. Entrepreneurial actors foster indigenous innovation capabilities by facilitating learning and upgrading the existing technological base with research institutions. University and research institution involvement associate with explicit intellectual property rights, thus, it provides convenience to commercialize technological knowledge via research knowledge spillovers (Hayter, 2016; Sternberg, 2014). Combining market knowledge with internal capabilities, entrepreneurial actors accumulate existing competence and commercialize fundamental research in the well-functioning market.

The dense networks between companies, investors, mentors, and other entrepreneurial actors enhance innovation efficiency, meanwhile, it forms a culture of reciprocity and encourages innovative and risk-taking behaviors to increase the prosperity of the entrepreneurial ecosystem. With the development of ZGC, Baidu, JingDong, and other successful companies present prominent histories of entrepreneurial success. It encourages entrepreneurial actors to contribute entrepreneurial efforts in the ecosystem. Thus, the well-functioning material attributes reinforce social and cultural attributes to expand and evolve the entrepreneurial ecosystem. Meanwhile, material attributes of ZGC continually evolve in corresponds to the emerging institutional context.

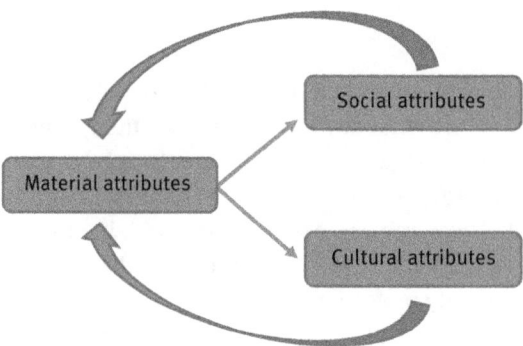

Figure 4.2: Relationship among entrepreneurial ecosystem in ZGC.

Zhangjiang High-Tech Park (ZJ), Shanghai

The emergence of the entrepreneurial ecosystem of ZJ is determined by regional re-sources and advantages. Institutional infrastructures and policies are inputs to stimulate the formation of the entrepreneurial ecosystem. Governments provide pol-icy levers to attract biomedical and integrated circuit foreign corporations and invest-ment. Meanwhile, ZJ promotes cooperation between foreign companies and local companies to reduce the risks associated with high technology tranfer and accelerate learning process. Diverting resources support the formation of related industrial chains and entrepreneurial ecosystems. Well-developed infrastructure and support-ive services become accessible in the entrepreneurial ecosystem (See Figure 4.3).

The entrepreneurial ecosystem serves as a conduit of novel ideas and knowledge, meanwhile, it assimilates external knowledge makes a greater contribution to inno-vation performance (Zahra & Nambisan, 2012). Material attributes exert on firms' strategies and capabilities in the entrepreneurial ecosystem to promote the invention and commercialization process. Entrepreneurs are passionate about new technolo-gies; thus, they form cooperative partnerships with external parties to invent and commercialize novel ideas (Shane & Venkataraman, 2000). Local companies engage in "reverse engineering" to transfer knowledge from cooperative partnerships. They assimilate and re-use external knowledge to continuously improve products and identify corporate competitiveness. Companies collectively redefine the specialization of the entrepreneurial ecosystem (Malecki, 2017). Transforming the dependencies on external parties in the ecosystem, local companies leverage internal capabilities and create new values. Assimilated knowledge provides the seeds for economic growth and prosperity. Companies in the entrepreneurial ecosystem gradually modify busi-ness models and develop dynamic capabilities to invent and commercialize innova-tion (Nambisan & Sawhney, 2008).

The formation of ZJ is based on entrepreneurial resource endowments, ZJ invests in R&D to promote breakthrough innovation and makes intensive capital investment in the pursuit of long-term market gains. This vision affects the propensity and appropriation of entrepreneurship. By promoting effective partnership with foreign global partners, local companies leverage and circulate resources to master the value chain and explore advantages in industrial production. Knowledge assimilation and innovation commercialization foster cooperative culture and catalysts the evolvement of the ecosystem.

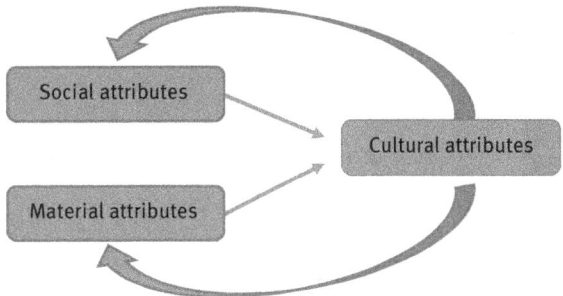

Figure 4.3: Relationship among entrepreneurial ecosystem in ZJ.

Nanshan Technology Park (NS), Shenzhen

Entrepreneurial ecosystems exhibit distinct features based on geographic and sociocultural characteristics (Roundy et al., 2018). Shenzhen is China's premier Special Economic Zone and becomes the fastest-growing city. Benefitting from friendly investment and economic practices, the NS thrives on the success in commercializing new technology and knowledge. Institutional policies result in the migration of talented human capital, extensive investment, and passionate entrepreneurs. Thus, it helps solidify an innovative and risk-taking culture that aggregates resources into the entrepreneurial ecosystem. The government is to cultivate the entrepreneurial community and culture through encouraging risk-taking and innovative behaviors. Moreover, the entrepreneurial culture is cultivated by enabling entrepreneurial actors to build a strong community and be tolerant of entrepreneurial failure. The recycling of talent and knowledge from failed experiences promotes the entrepreneurial culture. Entrepreneurial actors take innovative practices, accumulate knowledge and learn from failures (Cardon, Stevens & Potter, 2011).

Comparing with ZGC and ZJ, NS has limited research institutions. With limited spillover, companies in NS focus on innovation appropriation, including identifying promising applications, redefining existing markets and commercializing in the marketplace efficiently. The openness of interactions with corporate partners, knowledge is more fluid and organizational boundaries become transparent (See Figure 4.4).

NS presents greater resilience to respond to exogenous shocks and pressures (Cadenasso et al., 2006). Entrepreneurial actors maintain flexible business models, confine less bounded rationality, share risk-taking culture, and have greater capabilities to combine existing and emerging technologies (Shane & Venkataraman, 2000). The dominance of entrepreneurship and private enterprises brings resilience to transform its essential behaviors, structures, and identity into a system that is better able to respond to environmental turbulence (Walker et al., 2004). Leading companies integrate diverse resources to exploit their innovative solutions, meanwhile, entrepreneurial firms utilize complementary resources from leading companies to commercialize innovations in the huge domestic market. Thus, a resilient ecosystem can buffer environmental threats and appropriate complementary resources to facilitate commercial activities (Cenamor et al., 2013; Roundy et al., 2017).

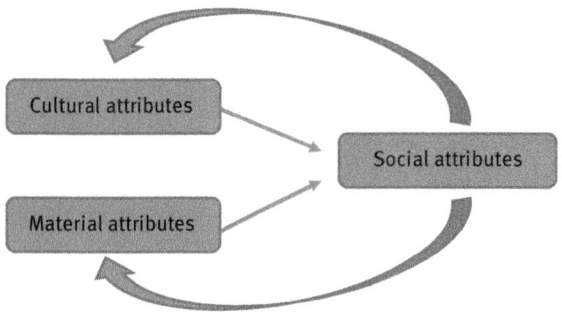

Figure 4.4: Relationship among entrepreneurial ecosystem in NS.

Overall, the dynamic interaction of resource attributes presents different configurations of the entrepreneurial ecosystem in China. The flourishing and self-evolving of the entrepreneurial ecosystem affects the innovation process (see Table 4.4). In ZGC, invention activities are carried out by universities and research institutions. SOEs and private enterprises accumulate existing competencies and commercialize fundamental research. The material attributes of ZGC drive social and cultural attributes to promote ingenious innovation capability. The preferential geographic advantage of ZJ attracts overseas technologies, talents and capital. ZJ makes the extensive investment to establish the IC industry value chain to assimilate external knowledge and commercialize innovation through cooperative partnerships. NS plays as the pioneer in economic reform, and it promotes entrepreneurial culture to encourage innovative and risk-taking behaviors. The shared entrepreneurial culture generates coherence to reinforce the resilience of the entrepreneurial ecosystem. Thus, entrepreneurial actors appropriate complementary resources from leading companies to commercialize innovation and strength the market position.

Table 4.4: Innovation process in entrepreneurial ecosystems.

	ZGC	ZJ	NS
The source innovation	Universities and research institutions	Foreign corporations	Local corporations
A set of activities drive the innovation process	Promoting basic research, and commercializing innovation through accumulated competencies	Assimilating external knowledge and commercializing innovation through cooperate partnership	Enhancing innovation efficiency and commercializing innovation through leveraging complementary resource
Lever of innovation process	Creative accumulation	Creative cooperation	Creative appropriation

Discussion and Conclusion

Implications for Entrepreneurial Ecosystems and Innovation Scholarship and Practice

The literature of the entrepreneurial ecosystem provides a holistic perspective to examine entrepreneurship and innovation. By examining entrepreneurial ecosystems in China, it has several contributions and implications for scholars and practitioners to explore future studies.

First, this paper draws attention to cultural, social and material attributes of the entrepreneurial ecosystem in China and identify the various configuration of ecosystem attributes. In ZGC, the strong set of material attributes helps to promote social and cultural attributes. With material attributes, ZJ's ecosystem is strengthening by social attributes, which attract foreign technology and investment to create opportunities for new ventures to access external knowledge. NS's ecosystem is driven by entrepreneurial culture, which encourages innovative activities. The well-functioning material attributes and underlying cultural attributes reinforce social attributes to make the entrepreneurial resource available for startups.

Second, this paper draws from the entrepreneurial ecosystem perspective to discuss how entrepreneurial ecosystems affect the innovation process. Different configurations of the entrepreneurial ecosystem produce and are reinforced by the innovation process. The dominance of material attributes cultivates indigenous innovative capabilities by exploiting breakthrough innovation from academic institutions and commercializing innovations. Creative accumulation is critical for corporations to invent and commercialize innovative products. Ecosystem oriented toward social attributes encourages cooperation to promote the innovation process, thus, creative coopera-tion is a critical force to assimilate and commercialize external knowledge. In addi-tion, ecosystems oriented toward cultural attributes present a greater tolerance and open environment for startups. Material and cultural attributes collectively aggregate market-oriented entrepreneurial resources and create social attributes. Creative ap-propriation occurs to leverage complementary resources to actualize entrepreneurial opportunities.

Conclusion

This paper examines the formation and evolvement of entrepreneurial ecosystems in China and acknowledges the variety of different configurations of the entrepreneurial ecosystem. Building on the theoretical framework, three entrepreneurial ecosystems are driven by different attributes and result in various opportunities and resources to affect the innovation process. Our aim is to draw attention to the innovation process and identify three kinds of innovative capabilities that would facilitate the process. The thriving entrepreneurial ecosystem depends on the configuration of innovative capabilities and attributes of the entrepreneurial ecosystem.

References

Acs, Z. J., Stam, E., Audretsch, D. B., & O'Connor, A. (2017). The lineages of the entrepreneurial ecosystem approach. *Small Business Economics*, 49(1), 1–10.
Adner, R. (2006). Match your innovation strategy to your innovation ecosystem. *Harvard Business Review*, April, 1–11.
Adner, R., & Kapoor, R. (2010). Value creation in innovation ecosystems: how the structure of technological interdependence affects firm performance in new technology generations. *Strategic Management Journal*, 31(3), 306–333.
Bahrami, H., & Evans, S. (1995). Flexible re-cycling and high-technology entrepreneurship. *California Management Review*, 37(3), 62–89.
Belling. (2021). *Company Profile*. Shanghai Belling. https://www.belling.com.cn/about.html.
Cadenasso, M. L., Pickett, S. T. A., & Grove, J. M. (2006). Dimensions of ecosystem complexity: heterogeneity, connectivity, and history. *Ecological Complexity*, 3(1), 1–12.
Cardon, M. S., Stevens, C. E., & Potter, D. R. (2011). Misfortunes or mistakes? *Journal of Business Venturing*, 26(1), 79–92.
Cenamor, J., Usero, B., & Fernández, Z. (2013). The role of complementary products on platform adoption: Evidence from the video console market. *Technovation*, *33*(12), 405–416.
CCIDNET. (2020). *A Report on 248 Unicorns in China*. CCIDNET. www.ccidnet.com/2020/0810/10537963.shtml.
Chen. (2003). *Chen Chunxian Studio*. Chenchunxian.com.http://chenchunxian.com/career.htm.
China Daily. (2019). *Zhangjiang sets up alliance to promote collaborative development of enterprises*. En.zjsfq.gov.cn.http://en.zjsfq.gov.cn/2019-07/17/c_388925.htm.

China Daily. (2020). *Zhongguancun Park*. Yooul.com.http://english.beijing.gov.cn/BeijingInfo/Sci/202005/t20200513_1896666.html.

Cohen, B. (2005). Sustainable valley entrepreneurial ecosystems. *Business Strategy and the Environment*, 15(1), 1–14.

Cohen, B. (2006). Sustainable valley entrepreneurial ecosystems. *Business Strategy and the Environment*, 15(1), 1–14

Cohen, S. L., Bingham, C. B., & Hallen, B. L. (2018). The role of accelerator designs in mitigating bounded rationality in New Ventures. *Administrative Science Quarterly*, 64(4), 810–854.

Dang, W., & Yan, S. (2020). *Nanshan District, the "Cradle" of Scientific and Technological Innovation Enterprises*. Baijiahao. https://baijiahao.baidu.com/s?id=1679141911026701379&wfr=spider&for=pc.

Dong, F. (2018). *Nanshan issued 1.42 billion yuan to support the development of enterprises*. lnanshan.sznews.com.http://inanshan.sznews.com/content/2018-05/31/content_21021609.htm.

Dougherty, D. (2004). Organizing practices in services: Capturing practice-based knowledge for innovation. *Strategic Organization*, 2(1), 35–64.

Dougherty, D., & Dunne, D. D. (2011). Organizing ecologies of complex innovation. *Organization Science*, 22(5), 1214–1223.

Eisenhardt, K. M., & Schoonhoven, C. B. (1996). Resource-based view of strategic alliance formation: strategic and social effects in entrepreneurial firms. *Organization Science*, 7(2), 136–150.

Fan, X. (2018). Research on the innovation environment of Shanghai enterprises, Based on the perspective of comparison with Beijing. *Journal of Shanghai Business School* (01), 77–86.

Feld, B. (2012). S*tartup Communities: Building an Entrepreneurial Ecosystem in Your City*. San Francisco: Wiley.

Gu, Y., & Sha, D. (2021). *Zhongguancun*: Science and technology embrace the market. *Big Science and Technology*, (1), 26–27.

Hayter, C. S. (2016). A trajectory of early-stage spinoff success: the role of knowledge intermediaries within an entrepreneurial university ecosystem. *Small Business Economics*, 47(3), 633–656.

He, W. (2019). *New Microsoft AI and IoT Research Lab to Commence Operations in May*. En.zjsfq.gov.cn.http://en.zjsfq.gov.cn/2019-04/12/c_354670.htm.

HLMC. (2020). *HLMC Corporate Profile*. Www.hlmc.cn.http://www.hlmc.cn/about_us.

Huahong Group. (2019). *Overview*. Huahonggrace.com.https://www.huahonggrace.com/html/business_over.php.

Isenberg, D. (2010). How to start an entrepreneurial revolution. *Harvard Business Review*, 88(6), 40–50.

Isenberg, D. (2011). *The Entrepreneurship Ecosystem Strategy as a New Paradigm for Economic Policy: Principles for Cultivating Entrepreneurship*. Institute of International and European Affairs.

Isenberg, D. (2011). The entrepreneurship ecosystem strategy as a new paradigm for economic policy: Principles for cultivating entrepreneurship. *Presentation at the Institute of International and European Affairs*, 1(781), 1–13.

Jing, S. (2019). Stock Connects Fueling Capital Flows via Mainland, HK Bourses. *China Daily*. http://www.chinadaily.com.cn/a/201911/19/WS5dd34016a310cf3e355783d1.html.

Kapoor, R., & Lee, J. M. (2012). Coordinating and competing in ecosystems: how organizational forms shape new technology investments. *Strategic Management Journal*, 34(3), 274–296.

Lei, A. (2018). *Report of the Development of Talent Innovation and Entrepreneurship in China*. Difang. https://difang.gmw.cn/gd/2018-12/02/content_32089807.htm.

Li. (2019). *Zhongguancun: A Banner of Innovation and Development*. Theory People. http://theory. people.com.cn/n1/2019/0111/c40531-30516210.html.

Liu. (2021). *Legend of Nanshan District: Shenzhen's GDP Accounts for 1/5 Trillion Years Later*. Money163. https://money.163.com/21/0126/01/G17U7BCA002580S6.html.

Ma, L., & Zeng, X. (2017). *The White Paper on the Development of Science and Technology Finance in Nanshan District*. INanshan. http://inanshan.sznews.com/content/2017-11/21/content_ 17794754.htm.

Malecki, E. J. (2018). Entrepreneurship and entrepreneurial ecosystems. *Geography compass*, 12(3), e12359.

Mason, C., & Brown, R. (2014). Entrepreneurial ecosystems and growth oriented entrepreneurship. *Final Report to OECD, Paris*, 30(1), 77–102.

Microwave Radio Network. (2004). *Comlent Technology Developed China's First Complete RF Integrated Circuit Chip*. MERF Net. http://www.mwrf.net/news/newtech/2004/14858.html.

Moore, J. (1993). Predators and prey: a new ecology of competition. *Harvard Business Review, May-June*, 75–86.

Roundy, P. T., Bradshaw, M., & Brockman, B. K. (2018). The emergence of entrepreneurial ecosystems: a complex adaptive systems approach. *Journal of Business Research*, 86(1), 1–10.

Roundy, P. T., Brockman, B. K., & Bradshaw, M. (2017). The resilience of entrepreneurial ecosystems. *Journal of Business Venturing Insights*, 8, 99–104.

Nambisan, S., & Sawhney, M. S. (2008). *The Global Brain: Your Roadmap for Innovating Faster and Smarter in a Networked World*. Philadelphia: Wharton School Pub.

Schumpeter, J. (1934). *The Theory of Economic Development*. Cambridge, Mass: Harvard University Press.

Shane, S., & Venkataraman, S. (2000). The promise of entrepreneurship as a field of research. *Academy of Management Review*, 25(1), 217–226.

Shenzhen Daily. (2020). *You Are a Shenzhener Once You Come Here*. Shenzhen Government. http://www.sz.gov.cn/en_szgov/news/photos/content/post_7900711.html.

Shenzhen Nanshan. (2013). *Park Profile of Shenzhen Nanshan Science and Technology Park*. Guangdong Net. https://gd.zhaoshang.net/yuanqu/detail/9765/intro.

SMIC. (2021). *About Us*. SMIC. http://www.smics.com/en/site/about_summary.

Spigel, B. (2015). The relational organization of entrepreneurial ecosystems. *Entrepreneurship Theory and Practice*, 41(1), 49–72.

Spigel, B., & Harrison, R. (2017). Toward a process theory of entrepreneurial ecosystems. *Strategic Entrepreneurship Journal*, 12(1), 151–168.

Stam, E. (2015). Entrepreneurial ecosystems and regional policy: A sympathetic critique. *European Planning Studies*, 23(9), 1759–1769.

Sternberg, R. (2014). Success factors of university-spin-offs: regional government support programs versus regional environment. *Technovation*, 34(3), 137–148.

Su, Y. (2019). *Zhangjiang: Driving a City's Transformation*. SHINE. https://www.shine.cn/biz/econ omy/1911085614.

Sun, & Jing. (2021). *Enterprises Play the Leading Role: The Secret of Shenzhen Innovation*. Baijiahao. https://baijiahao.baidu.com/s?id=1678441836529617078&wfr=spider&for=pc.

Tang, Y. (2015). *Shanghai: A New Hi-tech Center*. Bjreview. http://www.bjreview.com.cn/Nation/ 201509/t20150909_800037767.html.

Tansley, A. G. (1935). The use and abuse of vegetational concepts and terms. *Ecology*, 16(3), 284–307.

Telegent Systems. (2010). *Telegent Systems Produces of 100 million Mobile TV Chips*. Prnasia. https://www.prnasia.com/story/39237-1.shtml.

Theodoraki, C., Messeghem, K., & Rice, M. P. (2018). A social capital approach to the development of sustainable entrepreneurial ecosystems: an explorative study. *Small Business Economics*, 51(1), 153–170.

Walker, B., Holling, C. S., Carpenter, S. R., & Kinzig, A. (2004). Resilience, adaptability and transformability in social–ecological systems. *Ecology and Society*, 9(2).

Wang, Y. (2017). *Zhangjiang: Landlord to Shareholder*. East China. http://city.eastday.com/gk/20170413/u1a12888508.html.

Wang, Y. (2019). *Pudong to Get Major Role in Opening-up*. Zhangjiang Government. http://en.zjsfq.gov.cn/2019-06/26/c_384574.htm.

World Economic Forum. (2013). *Entrepreneurial ecosystems around the globe and company growth dynamics. Source: http://www3.weforum.org/docs/WEF_EntrepreneurialEcosystems_Report_2013.pdf (25.06.2020)*.

Welter, F. (2011). Contextualizing entrepreneurship-conceptual challenges and ways forward. *Entrepreneurship Theory and Practice*, 35(1), 165–184.

Xinhua. (2016). *Shenzhen Nanshan, Beijing Haidian, Shanghai Zhangjiang: Who will Become the "Silicon Valley of China"?* EET China. https://www.eet-china.com/news/201608051334.html.

Xinhua. (2019a). *Johnson & Johnson Innovation Launches Healthcare Incubator in Pudong*. Zhangjiang Government. http://en.zjsfq.gov.cn/2019-06/28/c_385296.htm.

Xinhua. (2019b). *Shanghai Sees Double-digit Growth in Foreign Investment*. Zhangjiang Government. http://en.zjsfq.gov.cn/2019-07/15/c_387999.htm.

Yang, R. (2018, April 11). *Believe It or Not – Shenzhen China Will Soon be Greater than Silicon Valley*. Dragon Social. https://www.dragonsocial.net/blog/shenzhen-silicon-valley/

Zahra, S. A., & Nambisan, S. (2012). Entrepreneurship and strategic thinking in business ecosystems. *Business Horizons*, 55(3), 219–229.

ZGC. (2021). *Zhongguancun Index 2020*. http://zgcgw.beijing.gov.cn/zgc/resource/cms/article/zgc_318/10910960/2020122409491273382.pdf.

ZGC Profile. (2021). *Zhongguancun National Independent Innovation Demonstration Zone*. Beijing Government. http://zgcgw.beijing.gov.cn/zgc/zwgk/sfqgk/sfqjs/fzlc/index.html.

Zhangjiang Profile. (2021). *Overview of Zhangjiang Science Park*. Pudong Government. http://www.pudong.gov.cn/shpd/gwh/023004/023004001/

Zhao, W., & Pira, F. L. (2013). Chinese entrepreneurship: institutions, ecosystems and growth limits. *Advances in Economics and Business*, 1(2), 72–88.

Zhuang, Tao. (2015). Resource integration perspective, study on the relationship between the three screw (Beijing university of posts and telecommunications). https://tra.oversea.cnki.net/KCMS/detail/detail.aspx?dbname=CDFDLAST2016&filename=1016015756.nh

Yumin Cao

5 Iteration of Digital Firms and Internationalization Theory: From the Perspectives of Speed and Learning

Introduction

What determines the internationalization performance of multinational corporations (MNCs)? From institutional and knowledge perspectives, internationalization speed has received attention in both traditional and emerging literature streams. By considering the digital aspect of internationalization performance, the transfer cost and arbitrage lead to potentially uncharted territory in internationalization theories, which requires further study.

Digital MNCs have been emerging as important players in international activities, especially the typical ones, born-global firms. Existing studies suggest two voices in digital MNC internationalization: intentional and accidental. The former is when MNCs undertake gradual global activities with previous internal planning, while the latter often refers to a company that internationalizes in a short time with little money. Born global literature focuses on the mechanism of the incredible speed of internationalization compared to the process of traditional MNCs, which could take decades. Given that digitalization plays an increasingly important role in this process, neither traditional international business theory nor born global studies explain the international expansion mechanism of digital MNCs.

This chapter is one of the few studies directly focusing on the mechanism of digital company internationalization, and we take an iterative perspective to answer this key question: what is the difference between traditional internationalization and digital internationalization? Specifically, we compare traditional IB theories, born global literature, and digital internationalization to study how the special operation process and iteration cycle influence the internationalization of born-digital firms.

Early theoretical models of internationalization emphasized the physical level of cross-border activities based on research on the gradual internationalization process of multinational corporations. They described the process of gradual internationalization by focusing on knowledge acquisition and experience accumulation; some examples are the Uppsala model and the OLI paradigm, among others (Johanson & Vahlne, 1977; Dunning, 1988), which concluded that a company's internationalization process is gradual. Hennart (2014) showed that, in addition to a company's intentional actions, the internationalization process may also be a result of a company's accidental behavior. Accidental internationalization refers to companies implementing new technologies to meet the needs of niche products with customers who are distributed globally and initially achieving high internationalization

https://doi.org/10.1515/9783110715002-005

speed through low transportation costs, sometimes even far beyond the company's expectations. We believe that the internationalization of born-digital firms is not completely gradual or accidental; instead, they use reasonable iteration cycles to obtain a large amount of market information and knowledge in a very short time at a low cost. With a unique balance between users, digital firms, and complementors in each host market, born-digital firms operate uniquely in different host markets through a continuous trial-and-error process.

In advancing this research, we show that born-digital firms' accidental internationalization is not the result of the company's intentions; on the contrary, it seems to be an accidental "reward" from the company's operations. Zeng et al. (2013) demonstrated that there are right and wrong experiences within acquisition; thus, the causality derived from such experiences can be right or wrong. Therefore, a company faces a process of "trial and error" to solve the internationalization of different cultural and institutional backgrounds, that is, an iterative process. We argue that the accidental internationalization of born-digital companies is not a pure accident but a process in which the company's trial-and-error process is so short that the "correct answer" is quickly found. In other words, the internationalization of born-digital companies is the result of their repeated iterative process to meet international market expectations. This study posits that the "accidental internationalization" of born-digital firms benefits from its flexible and short iterative cycle, and in host markets with long cultural and institutional distances, the impact of the liability of foreignness (LOF) can be reduced through such a process.

This chapter makes several contributions to the literature. First, based on traditional international business theory, we discuss the inexplicable and puzzling aspects of traditional perspectives on born-digital companies. Second, this chapter provides a new perspective for understanding the internationalization process of MNCs. Although we take digital enterprises as the study object, all multinational companies undergo constant learning and revision to acquire knowledge of the host country's market. The long iteration lengths of traditional companies may render the iterative process invisible and ignored. Third, based on the literature on rapid internationalization, this chapter expands our understanding of the mechanisms of rapid internationalization by delving into the mechanisms of the global expansion of digital firms. Rapid internationalization is not an accident, as previous studies have suggested, but an inevitable result. Fourth, we have practical implications for providing theoretical support for the rapid internationalization of digital MNCs. With the in-depth development of the information revolution, production and sales processes are increasingly using digital technology to enhance the competitiveness of enterprises in "going digital" or "gone digital" MNCs.

Traditional Internationalization Theory

Internalization and Externalization

Traditional international business theories emphasize that the internationalization of MNCs is a gradual process. A domestic company becomes a multinational player through the following steps: obtaining competitive advantages in the home country, exporting products to the host country, expanding its commitment in the host country, such as looking for agents, and, finally, establishing subsidiaries or acquiring local companies (Vahlne & Johanson, 1977). MNCs assess their advantages in terms of ownership, location, and internalization to decide on an international commitment (Dunning, 1988). This incremental internationalization theory is based on a key assumption: that the learning process and information acquisition of the company are continuous rather than a result of jumping through springboards; the company must accumulate enough knowledge to ensure the correctness of the information gathered before proceeding to the next step. This is because MNCs, as outsiders, often face high adaptation costs in unfamiliar local institutional and cultural environments, which is known as LOF (Zaheer, 1995). To overcome the LOF, multinational players should gradually gain unique knowledge and establish their reaction mechanisms in the host market. For example, traditional MNCs always look for an agent familiar with the local situation (O'Donnell, 2000), merge with a local company, or establish local subsidiaries with local employees to complete the production and transaction process locally (Vahlne & Johanson, 1977; 1990; 2017). In other words, through gradual internationalization, MNCs reduce the potential risks of cross-border activities over an elongated period. The company digests the adaptation costs and gains arbitrage space through its internal organizations and capabilities.

This gradual internationalization requires firms to internalize the host country's market knowledge to reduce transaction costs (Rugman, 1980; Dunning, 1988). Internalization refers to the failure of markets caused by the special nature of certain products or the existence of monopoly power, which leads to an increase in firms' market transaction costs (Buckley & Strange, 2011). Therefore, firms must internalize the external market into internal markets to ensure the flow of special knowledge within the enterprise and to reduce potential risks from information asymmetry (Doherty, 1999). However, increasing internalization increases the hierarchy cost in a vertical direction. Discussions of the benefits and hierarchy costs of internalization received early research attention in internalization studies (Narula, Asmussen, Chi, & Kundu, 2019). Hierarchy costs are endogenously related to experience, management skills, and conventions; therefore, companies vary in their ability to manage interdependencies within their corporate networks (Rugman & Verbeke, 1992). Intangible knowledge to effectively manage complex hierarchies is difficult to acquire and needs to be developed through experience. New MNCs that lack experiences may not have sufficient knowledge to achieve such internal coordination or effectively transfer firm-specific assets (FSAs) to different locations (Cuervo-Cazurra, 2012; Narula, 2017). The more complex the MNC's internal structure is, the more important it is (Birkinshaw & Pedersen, 2001). There is constant debate on promoting centralization and autonomy within MNCs (e.g., Young & Tavares, 2004).

In contrast, scholars have recently hypothesized that externalization (quasi-internalization) is another way to achieve international expansion of multinational enterprises. MNCs' organizational capabilities and sources of firm-specific resources can be balanced between internalization and externalization. In contrast to internalization, which emphasizes the vertical integration of external market knowledge and resources, externalization refers to the fact that MNCs do not pursue control over host country partners, implying inconsistencies between MNCs' control boundaries and ownership boundaries (Narula et al., 2019). It emphasizes that the involved parties are independent firms, but they promote mutual adaptation and reduce transaction costs through additional mechanisms, such as price, social connections, reputation and behavioral constraints (Hennart, 1993). MNCs no longer rely on internalization to reduce transaction costs. Instead, internal FSAs become labels and interfaces to the company's boundaries that externally link other partners. As a result, the acquisition of competitive advantages shifts from internalizing local knowledge to dealing with relationships between firms (Kedia & Mukherjee, 2009). The ability to build inter-company relationships facilitates cross-border FSA restructuring; as such, each company has its strengths, so participating parties can share the FSAs of others without fear of their partners taking advantage of them (Narula et al., 2015; 2019). Ultimately, MNCs and local partners gain the advantages of internalization but reduce the cost of internalization tiering. A typical example is global value chains, which treat externalization as steady cooperation, and scholars have attempted to understand the mechanisms that coordinate and control these externalizing activities (e.g., De Groot, Linders, Rietveld, & Subramanian, 2004; De Mendonça, Lirio, Braga, & Silva, 2014; Linders et al., 2005).

Born Globals and Born Digitals

In contrast to traditional MNCs, with the advancement of technology and diversification of information exchange channels, the rapid internationalization of companies has become an achievable option. This type of company operating international activities in a short period and whose goal is to become international at the beginning of its establishment are called born-global firms (Autio et al., 2000; McDougall & Oviatt, 2000; Knight & Cavusgil, 2004; 2005). Their appearance challenges traditional international business theories. For a born-global company, rapid internationalization means putting products in unfamiliar markets for a short period of time. Faced with relatively limited resources, firms proactively seek new opportunities in complex foreign markets that are often fraught with uncertainty and risk; thus, they may require an innovative, visionary, knowledge-gaining–oriented stance (Knight & Cavusgil, 2004).

Innovation plays an important role in global firms because the new knowledge brought about by innovation leads to the development of organizational structures, governance capabilities, and behavioral lines, which are the foundation for successful internationalization in foreign markets in the short term (Knight & Cavusgil, 2004). Innovation is the result of two pillars: knowledge within firms promoting R&D and the imitation of innovation by other firms (Lewin and Massini, 2004; Nelson and Winter, 1985). Therefore, to maintain a competitive advantage, companies must have unique, internalized knowledge to ensure that it is difficult to imitate and non-transferable. In essence, it is not the innovation itself but acting as a carrier of knowledge that brings special advantages to enterprises, which lays the foundation for rapid internationalization (Knight & Cavusgil, 2004; Kogut and Zander, 1993). New knowledge often leads to new paths that challenge current operating and management models, so the flexibility of young firms is stronger than that of older firms in terms of acquiring knowledge and innovation, and, ultimately, superior business performance is accomplished (Autio et al., 2000; Penrose, 1959). Specifically, knowledge about international markets and operations and the efficiency with which to acquire this knowledge are key determinants of successful international performance for born-global firms.

Given the accurate identification and utilization of knowledge, a global company's innovation capability is global and can be applied to various foreign markets in a short period of time through information technology and e-commerce capabilities (Zhang, & Tansuhaj, 2007). Unique innovative product development generates a differentiated strategic advantage in niche markets, thus minimizing direct confrontation with strong competitors (Knight & Cavusgil, 2004), resulting in position holding (Porter, 1980) and internationalized performance benefits. A firm's inimitable market-specific knowledge leads to high standard quality control, which increases customer loyalty and performance by promoting customer satisfaction. Finally, the market knowledge internalized by born-global firms is beneficial in finding skilled foreign intermediaries in host countries, building networks in a short period, and carrying out a range of local activities (e.g., Freeman, Edwards & Schroder, 2006; Thai & Chong, 2008).

Born-digital firms are the inheritance and further development of born-global firms, with reinforcement of information technology advantages. Lee et al. (2019) believe that digitalization has a positive moderating effect on a company's development of international capabilities. The company also needs to have traditional international business capabilities as well as the ability to control digitalization. The emergence of born digitals further expands the unknown areas of traditional international business theory. Although there is no unified understanding of the definition of born-digital firms, some characteristics have been recognized and widely accepted by scholars. The key feature is that the production, operation, and sales of born-digital enterprises are all dependent on digital technology such as networks (Monaghan,

Tippmann, & Coviello, 2020). These characteristics indicate two important conditions. One is the establishment and utilization of digital infrastructure to make the structure of the born-digital company a combination of digitization and substantiation (Nambisan, 2017; van Alstyne, Parker, & Choudary, 2016). Second, a company's products are digital and rely on digital technology for production and sales (Laudon & Laudon, 2015; Nambisan, 2017). Therefore, a distinctive feature between digital transformation companies and born-digital firms is that they take digital technology as their core technology and rely on it to establish production and sales models rather than enhancing the original production and sales process through digital technology (Monaghan et al., 2020). In short, the physical part of the company becomes a reinforcement for its digital part. As a result, when the born-digital enterprise internationalizes, the problem that the enterprise considers should no longer be investment but involvement. The local complementarity of resources is essential for value creation (Stallkamp & Schotter, 2021). Because of bounded rationality, traditional market transactions are unsustainable. For born-digital firms, local supplementary resources with which they should contract are unknown, and the solution to this market uncertainty relies on continuous learning after entry to acquire tacit knowledge. Moreover, involvement is more dynamic, which forces the involved parties' contracts to be in a state of constant "continuous improvement" because certain and completed contracts result in higher costs for all post-event adaptations. Therefore, neither externalization nor internalization alone can effectively explain the internationalization of born-digital firms. Details of comparison between born global companies and born digital companies will be presented in Table 5.1.

Similarly, innovation plays an important role in the internationalization of born-digital firms. Collaborative innovation from external partners, users, and the company meets the heterogeneous needs of various loosely coupled participants. Specifically, users reduce friction and barriers to communicating with each other through digital companies' technology (Ullah, Sepasgozar, & Wang, 2018; Chen et al., 2022), and they utilize user-generated content for advertising or third-party complementors' interactions to fulfill their interests (Sun & Zhu, 2013). Complementors use specific knowledge to develop unique innovative products to meet customer needs. The result of this multi-party interaction is that the division of labor among the participants is further refined, and the tacit knowledge of all parties continues to deepen so that the results of innovation become more efficient and difficult to imitate. Such a structure helps build multilateral insurance, on the one hand, by setting barriers for outside players; on the other hand, it protects the risk of internal imitation by all parties involved and ultimately increases the quantity and quality of the overall value creation. As recent research argues, the internationalization of digital companies moves beyond corporate-centric logic to the interaction of companies, user groups, and third-party complementors (Brouthers, Geisser, & Rothlauf, 2016; Chen et al., 2019; Coviello, Kano, & Liesch, 2017; Shaheer & Li, 2020; Li et al., 2019).

Table 5.1: Difference between born globals and born digitals.

	Born globals	Born-digitals
Structure	Firm as the center	Firm, user, complementor as a whole
Value creation	Unique products innovated by firm	Interaction and co-create by participants
Age & scale	Young and small	Young and small
Speed of internationalization	Quick (in 3years)	Immediate
Physical and digital relationship	Digital part enhances physical production	Physical part enhances digital operation

New Internationalization Model: Iteration Cycle

In a world of bounded rationality, individuals acquire information through past actions and experiences, which creates boundaries in their knowledge sets that constrain and shape their future transactions, whether within a company or among unaffiliated actors. Bounded rationality, therefore, limits the ability to build complete contracts, thus preventing participants from identifying potential risks and opportunistic behaviors (Narula et al., 2019). The limitation of information will cause multinational companies to fall into the uncertainty that "I do not know what I do not know"; such uncertainty rather than risk (Knight, 1921) may lead to the failure of various departments of the enterprise to maximize efficiency and ultimately lead to systematic market misjudgments (Langlois & Cosgel, 1993). Therefore, digital companies' internationalization may rely on local complementors (Li et al. 2019). At the same time, the digital code and user groups do not create value and bring financial performance by themselves; these unique factors should rely on external complementors to achieve their potential value. Digital companies refine their technology to attract users, and more users attract more complementors, creating a positive feedback loop. Therefore, although the internationalization of born-digital firms relies on multi-party interaction to achieve value creation for the firm itself, with the most important goals being ensuring the number of user groups and realizing the advantages of such user groups.

Potential digital product users will evaluate the platform's technology based on their perceptions, which means that overall, continuous functional improvements and added innovative features will more precisely meet user needs, resulting in a better quality of user perception. This may help digital enterprises to quickly occupy leading positions in the market. Interestingly, at the beginning of this process, born-digital companies are constrained by factors such as their small size and short establishment time, resulting in unknown customer needs in unfamiliar host markets. This "I don't know what I don't know" situation is similar to that of traditional multinational corporations. The lack of knowledge in digital companies stems from three sources. First, there is a lack of knowledge about potential complementors; therefore, resources cannot be obtained directly from complementors (Parker et al., 2016). Second, unfamiliar cultural and institutional environments pose challenges for digital companies to produce digital content, which will fail to meet user needs. Third, digital firms do not know in advance what resources and capabilities will be utilized (Furr & Shipilov, 2018).

These voids of knowledge lead to information on the users' side, and the information on the complementors' side may exceed the company's cognition; thus, although the digital company's assets may grow, the value created by the co-creation of the host country partners may decay from the host country (Zhang & Sarvary, 2015). Therefore, the internationalization of digital enterprises relies on reconfiguration after entering the host country rather than planning and preparing in advance. Essentially, it is the process of a continuous search for knowledge and new solutions to ensure that business activities continue to approach market needs. We call this trial-and-error improvement cycle the iteration cycle of born-digital firms. Iteration can be seen as an important part of the internationalization of a digital enterprise, as the right solution is not easy for new players to find and may require continuous subsequent time and investment. In other words, the host country's user group is not always what a born-digital firm wants it to be, and predicting everything will be difficult. For business managers, iteration is about continuous improvement, compromise, prioritization of what to improve and what not to improve, and the determination of what is adequate (rather than what is best) in the moment. It is a staged, multilayered process in which managers critically address what they have done to date in other markets.

Two Types of Iteration

First, iterations seek reasonable niche markets and build a differentiated advantage. In this type of iteration, born-digital firms use existing products or services without subversion or radical innovation. Born-digital firms entering foreign markets may hold limited explicit knowledge from information flows, and a lack of tacit knowledge builds barriers that prevent digital firms from distinguishing between potential users and actual users' preferences. Therefore, a reasonable and low-cost method is to use current mutual products or services entering the market and revising functions based on feedback from users. The original strategic goal can be achieved by constantly improving the host market. Given the heterogeneity of host market user characteristics, a successful digital product or service in host country A may not be able to repeat its achievements in host country B. Cultural differences may result in different or even opposite explanations of the same behavior or digital context. Institutional differences may result in higher legal adaptation costs for digital content regulations. Therefore, with an unchanged core business, born-digital firms can iterate to increase or decrease other services through the transformation from a single product/service to multiple products/services (or vice versa), which helps obtain a large local user group (as shown in Figure 5.1).

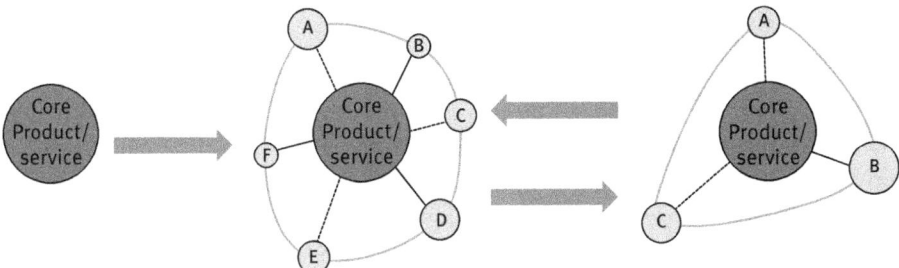

Figure 5.1: Product/service portfolio in different host markets.

Second, iteration involves seeking opportunities and taking low-cost risks. Popular products or services may differ by country (as shown in Figure 5.2) depending on customers' preferences, which are shaped by local cultural and institutional contexts. Therefore, operating a new digital product or service in foreign markets requires opportunity identification and risky behavior. Born-digital firms enjoy the newness advantages of flexibility resulting from their small scale, which makes this high trial-and-error cost acceptable through efficient communication, rapid decision-making, and relative objectivity of decision-makers. In contrast to old and large physical companies that fall into the "big enterprise disease," without path dependence or bureaucracy issues, born digitals are more able to transfer their core product or service at a low cost. The modularity and versatility of the technology transform core products/services independent of coordination among various internal stakeholders, and non-physical production makes transformation independent of the production line or raw material considerations.

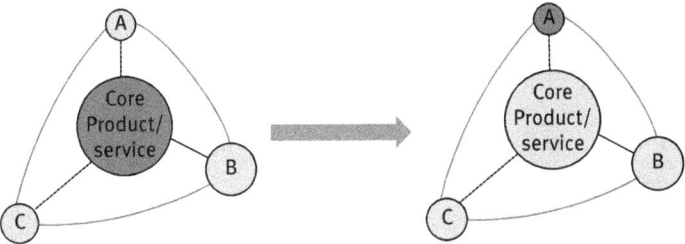

Figure 5.2: Core product/service transfer in different host markets.

Reasons for Quick Internationalization

Cost Transferring

The goal of digitization in a company is to drive out inefficient and high-cost parts that are internalized, including internal departments and external complementors. Therefore, the cost of internationalization has continuously passed from within the firm to local complementors, especially the actual economies in the host country. In the structure of users, digital enterprises, and complementors, the user group of the host country often does not need to pay the cost of using digital products, and the way digital companies achieve profitability is to rely on the user group to make profits through the complementors. Rapid internationalization means that companies find a suitable expansion model through iterations; in other words, after many trials and errors, they find a suitable "answer" for acquiring user groups. The question here is who bears the cost? After digital companies have established user-base advantages, local physical companies or other digital companies have become complementors of multinational companies. To market their products to these customers to gain market share, they are effectively in a low bargaining-power position in combination with digital multinationals. The iterative cost of information products is attached to the product or service, but it is eventually passed on to other local product producers or service providers. Of course, this is ultimately passed on to consumers in a broad sense. For example, most online shopping platforms in mobile app stores are free. Through interaction with the app, users can even buy physical products at prices far below the market price; for example, games and free online games are often filled with a large number of ads from physical industries and other digital companies. Therefore, the rapid internationalization of born-digital companies is not a miracle without cost but a process through which costs are passed on to other industries in the host country. Third-party complementors participating in value co-creation in the host country bear more risks and costs without gaining corresponding benefits. This multi-party interaction mechanism eventually forms not only within the digital enterprise but also outside the digital enterprise in which the digital part of the mechanism is strengthened by the physical part.

ESAs and Learning

Internationalization led by digital companies essentially builds multi-stakeholder value creation in the host country, which expands the learning process of previous multinational companies. Scholars have previously described the internationalization of a digital company as an ecosystem built in the host country, so the enterprise's firm-specific assets (FSA) are transformed into ecosystem-specific assets (ESA). In the absence of complementary assets, digital companies have limited independent advantages by relying solely on internalization. Conversely, one party may achieve higher/lower performance through the activities of the other participants (Jacobides et al., 2018). This is an important difference from the increasingly discussed quasi-internalization (externalization), which emphasizes the use of its own FSA as an interface to find external partners. The partners are independent of each other, and the focus is on how to use their own FSA. However, in the ecosystem of digital multinational enterprises, the various actors are in a state of loose coupling, and, more importantly, the process of turning FSA into benefits is no longer dominated by enterprises that internalize FSA but the result of synergy between each involved party. Given the distributed knowledge held by complementors, the point of innovation in the platform ecosystem is likely to lie outside the platform company; therefore, the constraints on the growth of the host ESA vary from place to place.

Digital information flow is rich in the system built by born-digital companies, which means that the threshold for obtaining explicit knowledge is further lowered. Although each complementor has independent knowledge, the transaction cost brought about by information asymmetry is reduced. Therefore, the important learning objective shifts from product/market to better coordinate the activities of external partners. Owing to the multilateral interdependence among ecosystem players and the increasing number of newly involved players, it is not easy to realign partners as the market evolves. To maintain the partnership and coordinate the activities of both parties, complementary companies often take the initiative to coordinate and simplify various activities by exchanging necessary information (Kulp et al., 2004; Monczka et al., 1998; Mohr et al., 1996; Mohr & Nevin, 1990; Mohr & Spekman, 1994; Cai et al., 2010). New product or technology information from managers of complementary firms is an important source of information, and frequent face-to-face interactions facilitate communication processes between firms (Kulp et al., 2004; Monczka et al., 1998) and enable information symmetry (Luk et al., 2008; Cai et al., 2010). The timely and minimal deviation information shared by complementary firms has the advantage of improving their survival and development ability (Luo, 2007).

An interesting phenomenon was observed in this study. The diverse information exchange channels created by digitalization have brought about great information flow, but they are not sufficient to meet the requirements of information symmetry. In contrast, digital enterprises must rely on non-digital informal information exchange channels to maintain information symmetry, albeit in a small proportion that is becoming increasingly important. Thus, under rich information flow, the learning of digital companies for relationship maintenance becomes particularly valuable, allowing digital companies to identify, in an iteration cycle, reasonable approaches to improve complementarity within the ecosystem. Ultimately, the value created by complementary externalities greatly exceeds transaction costs (Amit & Zott, 2001).

Moderate Innovation and Convergence

Radical innovation increases the cost of adaptation for users, and changes in modularity increase the cost of complementors, thereby detracting from the entire value co-creation process (Li, et al., 2019). The international value creation of born-digital companies does not come from their own FSA to gain a competitive advantage but through interactions with third-party complementors through the medium of user groups, so that the FSA of each participant serves the same purpose of interest. As a result, the over-investment of born-digital companies in their FSAs can hurt the overall architecture, as rapid technological capabilities (a key FSA for such companies) increase the development challenges of complementors and hinder their innovation, thereby compromising the value proposition of the entire ecosystem (Ozalp et al., 2018; Li et al., 2019). The introduction of new technologies can create discontinuities in older technologies and open windows of opportunity for competitors. New technological innovations, although they may have a similar technological upgrade path as do previous ones, also displace the value of older technologies, ultimately breaking established ties with their complementors.

This rich information flow brings about a reduction in barriers to explicit knowledge acquisition. Therefore, a company's successful model can be easily imitated by other companies, and similar products can be developed in a short period. Digital collaborators and competitors can share technology owing to their homogeneous application platforms and modular functions. Once new technologies or concepts appear, they are quickly internalized by digital enterprises in a short period. Ultimately, technology-sharing and moderate innovation lead to convergence among digital enterprises. For example, in terms of service functions, almost all digital enterprises have payment systems, social platforms, online shopping platforms, and so on. Similarly, after the success of short videos, almost all digital companies began to develop short video services within a short period of time. Therefore, digital enterprises cannot rely solely on a single market or a central market. The opening of new host markets has become the result of active demand and passive coercion.

Conclusion

Taking the perspective of learning and rapid internationalization, this chapter introduces a new type of company, the born-digital firm, and discusses the potential voids of traditional international business theory, especially in internalization and externalization. We study how the iteration cycle promotes the achievement of superior international activities in a short period of time. Pure internalization or externalization is insufficient to explain its global expansion, because users, digital firms, and complementors construct a complex and interactive structure that differs from host markets. Born-digital firms must offer unique solutions in each host market through a continuous trial-and-error process.

This study makes several contributions to the literature. First, it examines traditional international business and discusses its inexplicable part for born-digital firms. Traditional IB theory focuses on internalization or externalization by considering international expansion as a gradual process and emphasizing the core status of MNCs. This perspective shows voids in the digital era because advancing technology brings the possibility of rapid internationalization with deep cooperation in a short period.

Second, this chapter offers a new perspective for understanding the internationalization process. Although we use born-digital firms as a typical research object here, the iteration cycle, referring to continuous learning and revision to acquire host market knowledge, exists in all multinational companies. Owing to the length of the period for physical companies in this cycle, people ignore their existence. For both traditional MNCs and digital firms, learning from feedback is key to success; thus, the iteration cycle perspective offers a re-examination of the previously studied internationalization process.

Third, this chapter contributes to the rapid internationalization literature by deeply studying the mechanism of digital firms' global expansion. Born-globals have received wide attention from the literature, but few studies focus on the black box of the internationalization mechanism. The reasons we discussed in this chapter suggest that rapid internationalization is not an accidental action as previous studies have argued; it is an inevitable result.

Fourth, this study has important practical implications. Digitalization is the future development trend of multinational enterprises. Whether it is the "going digital" of traditional enterprises to digitalization or "gone digital" multinational enterprises, the use of digital technology is still increasing. This study provides theoretical support for the rapid internationalization of multinational companies in the context of digitalization.

Based on this internationalization process, future studies may further examine the direct impact of iteration on digital firms' performance or the indirect effect of the host country context on the iteration–performance relationship. For example, what is the difference between developing and developed countries when digital firms operate their iterative cycles? How does an informal information exchange channel influence the strategic decisions of digital firms in host markets? What is the cross-border impact of the cost transfer of digital firms in host markets? Future studies should consider these interesting directions.

References

Amit, R., & Zott, C. (2001). Value creation in e-business. *Strategic Management Journal*, 22(6–7), 493–520.

Autio, E., Sapienza, H. J., & Almeida, J. G. (2000). Effects of age at entry, knowledge intensity, and imitability on international growth. *Academy of Management Journal*, 43(5), 909–924.

Birkinshaw, J., & Pedersen, T. (2001). Strategy and management in MNE subsidiaries. In The *Oxford Handbook of International Business*. in A.M Rugman, T.L Brewer (Eds.), The Oxford handbook of international business, Oxford University Press, Oxford (2001), pp. 380–401

Brouthers, K. D., Geisser, K. D., & Rothlauf, F. (2016). Explaining the internationalization of ibusiness firms. *Journal of International Business Studies*, 47(5), 513–534.

Buckley, P. J., & Strange, R. (2011). The governance of the multinational enterprise: Insights from internalization theory. *Journal of Management Studies*, 48(2), 460–470.

Cai, S., Jun, M., & Yang, Z. (2010). Implementing supply chain information integration in China: The role of institutional forces and trust. *Journal of Operations Management*, 28(3), 257–268.

Chen, L., Li, S., Wei, J., & Yang, Y. (2022). Externalization in the platform economy: Social platforms and institutions. *Journal of International Business Studies*, https://doi.org/10.1057/s41267-022-00506-w

Chen, L., Shaheer, N., Yi, J., & Li, S. (2019). The international penetration of ibusiness firms: Network effects, liabilities of outsidership and country clout. *Journal of International Business Studies*, 50(2), 172–192.

Coviello, N., Kano, L., & Liesch, P. W. (2017). Adapting the Uppsala model to a modern world: Macro-context and micro-foundations. *Journal of International Business Studies*, 48(9), 1151–1164.

Cuervo-Cazurra, A. (2012). Extending theory by analyzing developing country multinational companies: Solving the Goldilocks debate. *Global Strategy Journal*, 2(3), 153–167.

De Groot, H. L., Linders, G. J., Rietveld, P., & Subramanian, U. (2004). The institutional determinants of bilateral trade patterns. *Kyklos*, 57(1), 103–123.

De Mendonça, T. G., Lirio, V. S., Braga, M. J., & Da Silva, O. M. (2014). Institutions and bilateral agricultural trade. *Procedia Economics and Finance*, 14, 164–172.

Doherty, A. M. (1999). Explaining international retailers' market entry mode strategy: Internalization theory, agency theory and the importance of information asymmetry. *The International Review of Retail, Distribution and Consumer Research*, 9(4), 379–402.

Dunning, J. H. (1988). The theory of international production. *The International Trade Journal*, 3(1), 21–66.

Freeman, S., Edwards, R., & Schroder, B. (2006). How smaller born-global firms use networks and alliances to overcome constraints to rapid internationalization. *Journal of international Marketing*, 14(3), 33–63.

Furr, N., & Shipilov, A. (2018). Building the right ecosystem for innovation. *MIT Sloan Management Review*, 59(4), 59–64.

Hennart, J. F. (1993). Explaining the swollen middle: Why most transactions are a mix of "market" and "hierarchy". *Organization Science*, 4(4), 529–547.

Hennart, J. F. (2014). The accidental internationalists: a theory of born globals. *Entrepreneurship Theory and Practice*, 38(1), 117–135.

Johanson, J., & Vahlne, J. E. (1977). The Internationalization Process of the Firm-A Model of Knowledge Development and Increasing Foreign Market Commitments. Journal of International Business Studies, 8(1), 23–32.

Johanson, J., & Vahlne, J. E. (2017). The internationalization process of the firm—a model of knowledge development and increasing foreign market commitments. In International business (pp. 145–154). Routledge.

Jacobides, M. G., Cennamo, C., & Gawer, A. (2018). Towards a theory of ecosystems. *Strategic Management Journal*, 39(8), 2255–2276.

Johanson, J., & Vahlne, J. E. (1990). The mechanism of internationalisation. International Marketing Review. 7(4), 11–24.

Kedia, B. L., & Mukherjee, D. (2009). Understanding offshoring: A research framework based on disintegration, location and externalization advantages. *Journal of World Business*, 44(3), 250–261.

Knight, F. H. (1921). *Risk, Uncertainty and Profit* (Vol. 31). New York: Houghton-Mifflin.

Knight, G., Cavusgil, S. T., & Innovation, O. C. (2004).The Born-global Firm. *Journal of International Business Studies*, 35(2), 124–141.

Knight, G. A., & Cavusgil, S. T. (2005). A taxonomy of born-global firms. MIR: *Management International Review*, 15–35

Kogut, B., & Zander, U. (1993). Knowledge of the firm and the evolutionary theory of the multinational corporation. *Journal of International Business Studies*, 24(4), 625–645.

Kulp, S. C., Lee, H. L., & Ofek, E. (2004). Manufacturer benefits from information integration with retail customers. *Management Science*, 50(4), 431–444.

Langlois, R. N., & Cosgel, M. M. (1993). Frank Knight on risk, uncertainty, and the firm: A new interpretation. *Economic Inquiry*, 31(3), 456–465.

Laudon, K. C., & Laudon, J. P. (2015). *Management Information Systems: Managing the Digital Firm*. Upper Saddle River, New Jersey: Prentice Hall.

Lee, Y. Y., Falahat, M., & Sia, B. K. (2019). Impact of digitalization on the speed of internationalization. *International Business Research*, 12(4), 1–11.

Lewin, A.Y., Massini, S. (2004). Knowledge Creation and Organizational Capabilities of Innovating and Imitating Firms. In: Tsoukas, H., Mylonopoulos, N. (eds) Organizations as Knowledge Systems. Palgrave Macmillan, London.

Li, J., Chen, L., Yi, J., Mao, J., & Liao, J. (2019). Ecosystem-specific advantages in international digital commerce. *Journal of International Business Studies*, 50(9), 1448–1463.

Linders, G. J., HL Slangen, A., De Groot, H. L., & Beugelsdijk, S. (2005). Cultural and institutional determinants of bilateral trade flows. *Tinbergen Institute Discussion Paper*, No. 05-074/3.

Luk, C. L., Yau, O. H., Sin, L. Y., Tse, A. C., Chow, R. P., & Lee, J. S. (2008). The effects of social capital and organizational innovativeness in different institutional contexts. *Journal of International Business Studies*, 39(4), 589–612.

Luo, Y., & Tung, R. L. (2007). International expansion of emerging market enterprises: A springboard perspective. *Journal of International Business Studies*, 38(4), 481–498.

Massini, S., Lewin, A. Y., & Greve, H. R. (2005). Innovators and imitators: Organizational reference groups and adoption of organizational routines. *Research Policy*, 34(10), 1550–1569.

McDougall, P. P., & Oviatt, B. M. (2000). International entrepreneurship: the intersection of two research paths. *Academy of Management Journal*, 43(5), 902–906.

Mohr, J. J., Fisher, R. J., & Nevin, J. R. (1996). Collaborative communication in interfirm relationships: moderating effects of integration and control. *Journal of Marketing*, 60(3), 103–115.

Mohr, J., & Nevin, J. R. (1990). Communication strategies in marketing channels: A theoretical perspective. *Journal of Marketing*, 54(4), 36–51.

Mohr, J., & Spekman, R. (1994). Characteristics of partnership success: partnership attributes, communication behavior, and conflict resolution techniques. *Strategic Management Journal*, 15(2), 135–152.

Monaghan, S., Tippmann, E., & Coviello, N. (2020). Born digitals: Thoughts on their internationalization and a research agenda. *Journal of International Business Studies*, 51(1), 11–22.

Monczka, R. M., Petersen, K. J., Handfield, R. B., & Ragatz, G. L. (1998). Success factors in strategic supplier alliances: the buying company perspective. *Decision Sciences*, 29(3), 553–577.

Nambisan, S. (2017). Digital entrepreneurship: Toward a digital technology perspective of entrepreneurship. *Entrepreneurship: Theory and Practice*, 41(6), 1029–1055.

Narula, R., & Verbeke, A. (2015). Making internalization theory good for practice: The essence of Alan Rugman's contributions to international business. *Journal of World Business*, 50(4), 612–622.

Narula, R., & Pineli, A. (2017). Multinational enterprises and economic development in host countries: what we know and what we don't know. In Development finance (pp. 147–188). Palgrave Macmillan, London.

Narula, R., Asmussen, C. G., Chi, T., & Kundu, S. K. (2019). Applying and advancing internalization theory: The multinational enterprise in the twenty-first century. *Journal of International Business Studies*, 50(8), 1231–1252.

Nelson, R. R. (1985). *An Evolutionary Theory of Economic Change*. Boston, MA: Harvard University Press.

O'Donnell, S. W. (2000). Managing foreign subsidiaries: Agents of headquarters, or an interdependent network? *Strategic Management Journal*, 21(5), 525–548.

Ozalp, H., Cennamo, C., & Gawer, A. (2018). Disruption in platform-based ecosystems. *Journal of Management Studies*, 55(7), 1203–1241.

Parker, G., Van Alstyne, M. W., & Jiang, X. (2016). Platform ecosystems: How developers invert the firm. Boston University Questrom School of Business Research Paper, (2861574).

Penrose, E. (1959) *The Theory of the Growth of the Firm*. New York: Oxford University Press.

Porter, M. E. (1980). Industry structure and competitive strategy: Keys to profitability. *Financial Analysts Journal*, 36(4), 30–41.

Rugman, A. M. (1980). Internalization as a general theory of foreign direct investment: A re-appraisal of the literature. *Weltwirtschaftliches Archiv*, (H. 2), 365–379.

Rugman, A. M., & Verbeke, A. (1992). A note on the transnational solution and the transaction cost theory of multinational strategic management. *Journal of International Business Studies*, 23(4), 761–771.

Shaheer, N., Li, S., & Priem, R. (2020). Revisiting location in a digital age: How can lead markets to accelerate the internationalization of mobile apps? *Journal of International Marketing*, 28(4), 21–40.

Stallkamp, M., & Schotter, A. P. (2021). Platforms without borders? The international strategies of digital platform firms. *Global Strategy Journal*, 11(1), 58–80.

Sun, M., & Zhu, F. (2013). Ad revenue and content commercialization: Evidence from blogs. *Management Science*, 59(10), 2314–2331.

Thai, M. T. T., & Chong, L. C. (2008). Born-global: The case of four Vietnamese SMEs. *Journal of International Entrepreneurship*, 6(2), 72–100.

Ullah, F., Sepasgozar, S. M., & Wang, C. (2018). A systematic review of smart real estate technology: Drivers of, and barriers to, the use of digital disruptive technologies and online platforms. *Sustainability*, 10(9), 3142.

van Alstyne, M. W., Parker, G. G., & Choudary, S. P. 2016. Pipelines, platforms, and the new rules of strategy. *Harvard Business Review*, 94(4): 54–62.

Young, S., & Tavares, A. T. (2004). Centralization and autonomy: Back to the future. *International Business Review*, 13(2), 215–237.

Zaheer, S. (1995). Overcoming the liability of foreignness. *Academy of Management Journal*, 38(2), 341–363.

Zhang, K., & Sarvary, M. (2015). Differentiation with user-generated content. *Management Science*, 61(4), 898–914.

Zhang, M., & Tansuhaj, P. S. (2007). Organizational culture, information technology capability, and performance: the case of born global firms. *Multinational Business Review*, 15(3), 43–78.

Zeng, Y., Shenkar, O., Lee, S. H., & Song, S. (2013). Cultural differences, MNE learning abilities, and the effect of experience on subsidiary mortality in a dissimilar culture: Evidence from Korean MNEs. *Journal of International Business Studies*, 44(1), 42–65.

Lancy Mac and Felicitas Evangelista

6 Can Organizational Learning Makes Chinese Exporting Firms More Entrepreneurial?

Introduction

International businesses face highly complex environment as their operations span multiple national boundaries. When firms internationalize, the requirements on their skills and capabilities multiply. Being entrepreneurial is one among the numerous facets of a firm's capabilities that is deemed to be essential. In fact, internationalizing is already an entrepreneurial activity (Casson, 2000; Schumpeter, 1939; Simmonde & Smith, 1968) as engaging in business across international borders requires adaptation of otherwise usual practices as well as a certain amount of risk taking. Thus, entrepreneurship is of crucial importance to international performance and is actually becoming a necessity (Covin & Slevin, 1989; Lumpkin & Dess, 1996). Yet, research investigating this relationship is still scarce (Balabanis & Katsikea, 2003; Lampe et al., 2020) although those in the non-international settings are ample (Barringer & Bluedorn, 1999; Colvin & Slevin, 1991; Zahra, 1991, 1993, to name just a few).

Organizational learning has long been postulated to be an essential antecedent of performance with ample empirical evidence (Slater & Narver, 1995; Baker & Sinkula, 1999). It is also among the most important factor in achieving a sustainable competitive advantage (Cohen & Levinthal, 1990; Kogut & Zander, 1992) as collective learning by employees is difficult to imitate so it is important to stimulate learning among employees (de Geus, 1988, Sinkula et al., 1997; Hamel & Prahalad, 1994). Organizational learning is regarded as particularly essential during international expansion due to the "liability of foreigness" whereby the company is unfamiliar with the foreign market, customers and environment (Zaheer, 1995).

This study argues that both entrepreneurship and learning are important contributors of firm performance for exporters and that both can and should co-exist to contribute to desirable outcomes. The rationale behind such proposition is that when exporters seek opportunities abroad, they would need to learn about the new market. What works in their home country may not do so in the host countries, so substantial adaptation in terms of products, processes or systems will have to be done. When crossing national boundaries, they would also need to be adventurous and daring – in other words being entrepreneurial. Both of these are essential in ensuring successful performance in international markets. The aim of the current study is to investigate the relationship between entrepreneurship and organizational learning and their effects on export performance.

https://doi.org/10.1515/9783110715002-006

Specifically, we argue that organizational learning can enhance international corporate entrepreneurship (ICE) which in turn leads to firm performance. Organizational learning involves developing new knowledge or insights that can potentially influence behavior which can improve performance (Fiol & Lyles, 1985; Huber, 1991). These new knowledge is able to help update the company's usual practice and therefore generate new ways of doing things or even new products/services. Therefore, learning plays a key role in enhancing entrepreneurship of firms.

This paper seeks to contribute to the literature of entrepreneurial organizations (Lampe et al., 2020) in facilitating understanding of how entrepreneurship is managed in an international context. It also seeks to integrate learning and entrepreneurship as two distinct yet related research areas. As such this can help to address the research void in the literature which lacks studies on the role of entrepreneurship in the international context as well as its integration with organizational learning.

This study is undertaken in the context of the transitional economy of the People's Republic of China. Being the largest exporter in the world (World Bank, 2020), the large number of exporters will need to be equipped with various capabilities in order to compete in the highly competitive global marketplace. It would be imperative to investigate how exporters from emerging economies can cultivate the necessary capabilities to compete effectively in global markets.

Conceptual Model and Hypotheses

Entrepreneurship

The importance of entrepreneurship is well documented in the literature, noticeably since the publication of Drucker (1954) who asserted that there are only two basic functions of a business: marketing and innovation, where innovation is a key integral part of entrepreneurship (Limpkin & Dess, 1996; Lampe et al., 2020). It has been heavily researched yet with different labels including entrepreneurship (Miller 1983), intrapreneurship (Kuratko, Montagno & Hornsby, 1990), corporate entrepreneurship (Peterson & Berger, 1971; Morris & Paul, 1987; Zahra, 1991; Zahra & Covin, 1995), entrepreneurship posture (Covin & Slevin, 1991; Balabanis & Katsikea, 2003), entrepreneurial proclivity (Pellisier & Van Buer, 1996; Matsuno, Mentzer & Ozsomer, 2002), entrepreneurial orientation (Lumpkin & Dess, 1996), international corporate entrepreneurship (Zahra & Garvis, 2000), international entrepreneurship (McDougall, 1989) etc.

Schumpeter (1950, p.72) stated that "the function of entrepreneurship is to reform or revolutionize the pattern of production". Later, Burgelman (1984, p.154) defined it as the "process of extending the firms' domain of competence and corresponding opportunity set through internally generated new resource combinations". While some studies refer the term exclusively to the establishment of new business ventures (Garvin & Levesque, 2006), most identify it as the process of organizational renewal (Sathe, 1989). The latter suggests that it does not only encompass setting up new ventures or developing new product but also include innovations in services, channels, brands (Sawhney, Wolcott & Arroniz, 2006), processes and market developments (Zahra, 1991). Despite the differences in conceptualization, entrepreneurship is regarded as a strategy for coping effectively with competition in world markets (Kuratko et al., 2015). Studies have shown that corporate entrepreneurship could impact a company's competitiveness, its agility and ability to engage in continuous innovation and renewal (Urbano et al., 2022).

"Corporate entrepreneurship" is often used in the context of established organizations while "international entrepreneurship" is used when an international context is involved. As the main purpose of this study is to investigate firm-level entrepreneurial behavior of exporters, the construct *international corporate entrepreneurship* (ICE) will be adopted.

International Corporate Entrepreneurship (ICE) and Export Performance

Entrepreneurship is of crucial importance to companies expanding abroad as foreign markets tend to be more complex and unpredictable. The ability to innovate and take risk can readily prepare firms to adjust to the new environment and thus foster performance in the foreign market. With the seminal article of McDougall (1989) which compare the differences between domestic and international entrepreneurship, this concept is being applied in the context of international business. Initially, entrepreneurship was studied to explain how born global firms, those that internationalized at an early stage of inception and derived most of their revenue from abroad, were able to achieve success. Gradually, pleas were made to expand its domain (Giamartino et al., 1993) to include new and established companies as well as large and small firms (Stevenson & Jarillo, 1990; Covin & Slevin, 1991; Naman & Slevin, 1993) as it is argued that all firms need to be entrepreneurial with the increasingly hostile environment. Nonetheless, the amount of scholarly work devoted to international corporate entrepreneurship according to Glinyanova et al. (2021) remains scarce.

Dimitratos and Plakoyiannaki (2003, p.189) defined international entrepreneurship as "an organization wide process which is embedded in the organizational culture of the firm and which seeks through the exploitation of opportunities in the international marketplace to generate value". In other words, it is an organizational value and belief that can enhance better grasps of opportunities when moving to a different market. Similarly McDougall and Oviatt (2000, p.903) defined it as "combination of innovative, proactive, and risk taking behavior that crosses national borders and is intend to create value in organizations". Therefore, international corporate entrepreneurship can simply be understood as the propensity to innovate, take risk and act before competitors do when crossing international boundaries.

The effect of corporate entrepreneurship on performance is well documented in literature with a large number of studies providing empirical evidence for their positive relationship (e.g. Qiuqin et al. 2020; Bierwerth et al. 2015; Barringer & Bluedorn, 1999; Zahra, Jennings & Kuratko, 1999; Kropp, Lindsay & Shoham, 2006; Matsuno et al., 2002). Studies can also be found in the international context like Zahra and Garvis (2000), Balabanis and Katsikea (2003), Liu, Li and Xue, 2011 and Naldi et al. (2015). It is not difficult to imagine that exporters are faced with various challenges in foreign markets due to their "distance" from their customers abroad. Oftentimes, these exporters have to predict what their customers want and be innovative in their actions, all of which involve significant risk-taking. It is therefore hypothesized that:

H1: International corporate entrepreneurship is positively related with export performance

Organizational Learning and International Corporate Entrepreneurship

There has been a growing interest in organizational learning (Senge, 1990; Day, 1991; Galer & van der Heijden, 1992; Tobin, 1993; Garvin, 1993; Moorman, 1995) because of its important role in stimulating organizational performance (Slater & Narver, 1995; Baker & Sinkula, 1999) as well as sustainable competitive advantage (Slater & Narver, 1994, 1995). Organizational learning is the development of new knowledge or insights that can potentially influence behavior which can improve performance (Fiol & Lyles, 1985; Huber, 1991). It is the capacity to improve performance based on past experience (Nevis, DiBella & Gould, 1995), ability to infer from past endeavors in guiding future behavior (Levitt & March, 1988) and to engage in systematic problem solving and experimentation (Garvin, 1993). Through these, the cognitive maps of the firm, markets and competitors can be reshaped (De Geus, 1988), errors be detected and corrected (Argyris & Schon, 1978) and actions be improved based on better knowledge and understanding (Fiol & Lyles, 1985). According to Garvin (1993, p.80), a learning organization is one which is "skilled at creating, acquiring and transferring knowledge and at modifying its behavior to reflect new knowledge and insights". In other words, learning can enable firms to rapidly acquire knowledge and transform themselves. However, being a learning organization is not enough; it should be able to translate learned processes into managerial competence allowing it to serve customer needs more effectively (Hamel & Prahalad, 1994).

Learning and entrepreneurship are both distinct yet related concepts. Learning in organizations may take many forms and a common approach is by trial-and-error, that is, those behaviors that will lead to positive outcome will be repeated while those that lead to negative outcome will be avoided (Levitt & March, 1988). In the words of Lant and colleagues, learning is a discovery of discrepancies (Lant, 1989) and new ways of doing things (Lant & Mezias, 1992). Entrepreneurship also involves substantial searching of alternatives like what learning does. The main objective of searching is to adjust organizational behavior to discover possibility of change in the organization. As Kirzner (1979) suggests, such search for better alternatives (products or processes) is highly uncertain and a lot of resources have to be deployed yet may not ensure desirable outcome. Oftentimes, it will require utilizing resources in new ways or even create new resources (Zahra, Jennings & Kuratko, 1999). Corporate entrepreneurship allows values be created with reshuffling of resources and also create knowledge and knowledge-based outcomes – something that learning can also create (Zahra et al., 1999). Thus, both entrepreneurship and learning can help to produce the necessary knowledge base of the firm. The high connectivity of the two constructs can be reflected in the emergence of research on "entrepreneurial learning" (Minniti & Bygrave, 2001; Politis, 2005; Rupcic, 2019).

The integration of learning and entrepreneurship within organizations is quite promising and increasingly more studies can be found. Liu, Luo and Shi (2002) undertook a study of the effect of corporate entrepreneurship on learning orientation of state-owned enterprises in China. On the contrary, Zehir and Eren (2007) investigated the effect of learning orientation on corporate entrepreneurship in the Turkish automotive industry and found that a learning orientation contributes to all aspects of corporate entrepreneurship including new business venturing, innovativeness, proactiveness and risk taking. Similarly, Hult and Kandemir (2003) found that learning orientation leads to innovativeness and this relationship is moderated by global organizational memory. More recently, Wolff et al. (2015) also found a contributive effect of learning orientation on entrepreneurial orientation among small firms. In the international context, the scarce empirical evidence seems to suggest that entrepreneurship is related to organizational learning capability (e.g. Fernandez-Mesa & Alegre, 2015). There appears to be a gap in the literature which this study seeks to fill.

Organizational learning can be manifested in different forms and in this study we propose using the organizational value towards learning (learning commitment), the way of learning (learning style) as well as the capability to learn (absorptive capacity).

Commitment to Learning

Commitment to learning refers to the fundamental value a firm holds towards learning (Sinkula et al., 1997). This value determines the learning culture of an organization. It is associated with a firm's learning principles (Senge, 1990), ability to think and reason (Tobin, 1993) and ability to understand the firm's environment (Galer & van der Heijden, 1992). A firm that is committed to learning also values the need to understand the cause and effects of their actions which is necessary to assess and revise the theory in use (Shaw & Perkins, 1991). Without commitment, motivation to learn would be low and little learning is likely to occur (Sackmann, 1991). For these reasons, commitment to learning is considered as a key ingredient in instilling and facilitating a learning orientation in an organization (Slater & Narver, 1995).

Learning orientation has been associated with corporate entrepreneurship in previous studies (e.g. Dess et al., 2003; Liu, Luo & Shi, 2002). The view held in these studies is that corporate entrepreneurship leads to new knowledge which in turn affects performance. Another view is that it is through organizational learning that firms become aware of changes in the external environment causing them to engage in corporate entrepreneurial activities in order to adapt and remain competitive (Sambrook & Roberts, 2005; Wolff et al., 2015). In the current study, the latter view is adopted as it can be argued that commitment to learning is what drives firms to learn and become innovative, proactive and be more confident in dealing with their environment. An organization committed to learning is also more likely to create an internal environment that encourages employees to take personal initiative in pursuing innovative activities. Such internal environment also instils innovativeness as shown by a study conducted by Calantone, Cavusgil, and Zhao (2002) who found a clear link between learning and corporate entrepreneurship. Applying the above arguments to an international context, the following hypothesis is formed:

H2: Commitment to learning is positively related with international corporate entrepreneurship.

Learning Style

The aim of organizational learning is to enhance the competence of both employees and the entire organization. The learning style or system that an organization adopts can determine how well this aim is achieved. Two learning systems are commonly cited in the literature: single and double loop learning. Single loop learning is concerned with the utilization of existing knowledge and experience in solving problems and improving performance. The process of learning is in terms of detecting errors and making corrections without changing the established set of governing variables. Double loop learning, on the other hand, occurs when the underlying assumptions of the governing variables are questioned and challenged (Argyris & Schon, 1978).

Single loop learning may be viewed as low order learning while double loop as high order learning. The former is adopted by companies dealing with stable markets and when customers are seeking standard rather than customized products or services. It involves "doing things better" while the latter is likened to "doing things differently" (Hayes & Allinson, 1998). Single loop learning is also said to have similarities with adaptive behavior while double loop learning with innovative behavior (Korth, 2000). Higher order or double loop learning would tend to be generative rather than adaptive. Because of its nature, double loop learning allows the firm to interpret and identify emerging trends as well as drawing non-obvious conclusions (Lei et al., 1996). Firms engaged in higher-level learning are therefore more likely to create or identify opportunities faster than others so be able to yield better returns (Alvarez & Busenitz, 2001). As such, it would help management look ahead and identify opportunities (Day, 1994). The relationship between learning style and international corporate entrepreneurship is thus hypothesized as follows:

H3: There is a positive relationship between learning style and international corporate entrepreneurship.

Absorptive Capacity

The ability to innovate depends not only on how we learn but also on how much we know. To be able to absorb new knowledge, it is essential that one should possess relevant prior knowledge. Cohen and Levinthal (1990) suggested that prior related knowledge allows one "to recognize the value of new information, assimilate it and apply it" (p. 128), which is referred to as absorptive capacity. External information are abundant and they may be of little value if they are not seen to be useful. Managers with absorptive capacity are able to see what is relevant and important and these information got distilled into the organization so that emerging opportunities can be capitalized in a timely manner through an entrepreneurial orientation (Augier et al., 2001). In dynamic environments, opportunities tend to be short-lived so it is crucial that managers are able to move quickly to "absorb" these emerging opportunities (van Doorn et al., 2017). It has also been shown that the combination of present and past knowledge can lead to more effective learning (Cohen & Levinthal, 1990) and that learning performance will be enhanced when the object of learning is related to what is already known (Inkpen, 2000). Similarly, Powell et al. (1996) argue that knowledge facilitates the use of other knowledge and what can be learned is crucially affected by what is already known. As expressed by Brockmann and Anthony (2002, p.439), "the more we know, the more we can learn".

Previous studies have identified the skills of the export team as indicator of absorptive capacity which has a significant influence on the international operations of a firm. The complexities and uncertainties of international environments call for a capable team that can process information in order to recognize opportunities and resolve problems. One characteristic deemed important is the team members' international experience. Managers with previous international experience have been found to pursue international markets faster (Autio et al., 2000). They are found to have greater awareness of emergent opportunities in foreign markets (Westhead et al., 2001) and to perceive foreign markets to be less complex than their less experienced counterparts (Johansson & Vahlne, 1977). Aside from international experience, previous industry experience and educational background are also found to be associated with entrepreneurial ventures (Evangelista, 2005). In effect, previous industry and international experience, and educational background would indicate the learning capability of the export team. A team that is highly capable of learning would be able to take an entrepreneurial approach to changes in the firm's environment. Through continuous learning and accumulated knowledge they would be able to detect opportunities, the uncertainty of operating abroad would be reduced, perceived foreign expansion cost would be lower and commitments to foreign ventures would be greater (Johanson & Vahlne, 1990). On this basis, the following hypothesis is formed:

H4: The learning capability of the export team is positively related with international corporate entrepreneurship.

The relationship among the variables is diagrammatically depicted in Figure 6.1.

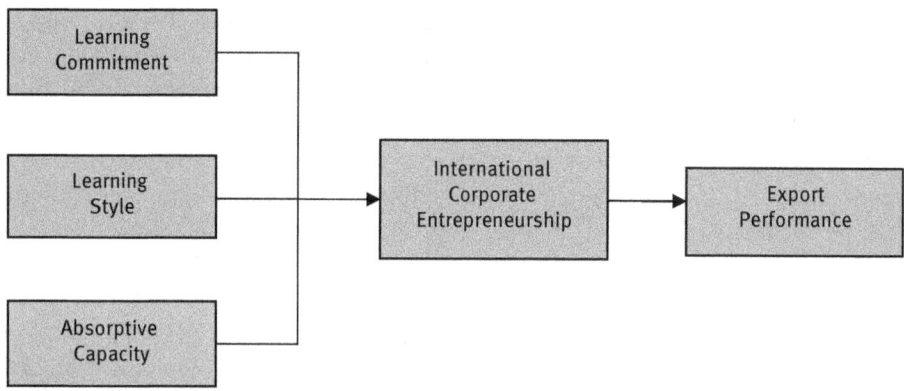

Figure 6.1: Conceptual model.

Methods

Data Collection

Personal interviews using a structured questionnaire were undertaken to collect data from exporters in the two provinces in South China, namely Guangdong and Fujian. These two provinces are the main exporting provinces in China (Xinhua

Net, 2015). Data were collected from the manager identified to be responsible for the exporting operation of the sample firms. In some cases, this may be the marketing manager, export manager or the general manager. A total sample of 128 firms were interviewed. Most of them are private companies (53.1%), are small and medium sized (79.4%) and mainly export consumer products (66.7%).

Measures

All scales used were adapted from previous studies. International corporate entrepreneurship (ICE) is measured by a seven-item scale taken from Zahra and Garvis (2000)'s study who modified the original corporate entrepreneurship scale of Miller (1983). The latter was used extensively in the corporate entrepreneurship literature and tested to be both valid and reliable. The scale for commitment to learning is adopted from Sinkula et al (1997) while that for learning style is from Chaston, Badger and Sadler-Smith (2000)'s study. The scale for absorptive capacity is adapted from Le and Evangelista (2006)'s scales measuring export team's learning capacity. Export performance is measured by subjective evaluation, a method deemed to be appropriate and desirable for studying emerging businesses (Chandler & Hanks, 1993). All scales are measured on 7-point scale except for ICE which is measured on a 5-point scale. Two control variables were included namely the type of industry and ownership. Both of them are dummy coded with 1 representing that they belong to the service industry and are indigenous companies with local capital, respectively.

Results

Scale Validation

Measures of the constructs were subjected to confirmatory factor analysis (CFA) to test their convergent and discriminant validity. Due to the small sample size, two CFAs were conducted. For ICE and learning commitment, inspection of the standardized loadings suggested that two items in the ICE scale were candidates for deletion due to their low standardized loadings (below 0.5). After removing these two items, the fit indices were chi-square=42.436, df=19, p=0.002; RMR=0.066; GFI=0.923; CFI=0.949. Learning style and absorptive capacity were subjected to another CFA and 4 items were removed from the learning style scale with the resulting fit indices being chi-square=51.505; df=19, p=0.000; RMR=0.089; GFI=0.919; CFI=0.929. Table 6.1 shows the standardized loadings from the CFAs, as well as the composite reliability and average variance extracted (AVE) while Table 6.2 shows the correlation coefficients. All

Table 6.1: Standardized loadings of CFAs, composite reliability and average variance Extracted.

Scales	Standardized loadings	Composite reliability	AVE
ICE		0.795	0.541
1. Shows a great deal of tolerance for high risk projects	0.739		
2. Uses only "tried and true" procedures, systems and methods*	–		
3. Challenges rather than responds to its major competitors	0.558		
4. Takes bold, wide-ranging strategic actions, rather than minor changes in tactics	0.659		
5. Emphasizes the pursuit of long-term goals and strategies*	–		
6. Is the first in the industry to introduce new products to the market	0.598		
7. Rewards taking calculated risks	0.742		
Learning commitment		0.891	0.761
1. Our company's ability to learn is the key to our competitive advantage.	0.808		
2. The basic values of this company include learning as a key to improvement.	0.918		
3. The sense around here is that learning is an investment, not an expense.	0.839		
Learning style		0.869	0.571
1. Constructive feedback is given to employees on how they are doing	0.635		
2. Employees are encouraged to pursue higher educational degrees	0.788		
3. Employees are encouraged to undertake training and development	0.812		
4. Employees share training lessons with others	0.816		
5. Company goals are made clear to all employees	0.717		
6. Employees, suppliers, customers are encouraged to let company know if anything is going wrong*	–		
7. Employees are not afraid to voice differing opinions*	–		
8. Company is always willing to change working practices*	–		
9. Company is on the lookout for new ideas from any source*	–		
Absorptive capacity		0.786	0.553
1. Our export team consists of members who have previous international experience	0.727		
2. Our export team has previous industry experience	0.837		
3. Our export team has relevant formal training or University degree.	0.656		

*Items deleted.

composite reliabilities are above the recommended threshold of 0.7 and AVE above 0.5 indicating good convergent and discriminant validity (Fornell & Larcker, 1981).

To ensure that the results are free from common method bias, one scale used a 5-point Likert scale (ICE) while other variables utilized a 7-point scale in order to create a psychological separation that can deter respondents from answering in a similar manner (Podsakoff et al. 2003). Harman's one-factor approach (Podsakoff & Organ, 1986) was also used. Four factors emerged with the first factor explaining only 41.8% of the total variance indicating that common method variance is not an issue.

Table 6.2: Descriptive statistics and correlations.

	Mean	S.D.	2	3	4	5
1. Learning style	5.463	0.872	.527**	.605**	.590**	0.120
2. Absorptive cap	5.449	1.073		.338**	.296**	.204*
3. Learning commitment	5.215	1.369			.598**	.218*
4. ICE	3.498	0.734				.419**
5. Export performance	5.351	1.430				

**p<0.01, *p<0.05

Hypothesis Testing Results

Structural equation modelling (SEM) was performed using AMOS to test the hypothesized relationships. Results show reasonably good fit indices: chi-square=24.283, df=12, p=0.019; RMR=0.062; GFI=0.949; CFI=0.938. Learning style and learning commitment are positively related to ICE, while absorptive capacity is not. ICE is found to be positively related to export performance. Therefore, H1, H2 and H3 are supported while H4 is not. Results are summarized in Table 6.3. None of the control variables is found to have a significant relationship.

Table 6.3: SEM results.

	Standardized beta	Hypothesis testing results
H1: ICE --> Export performance	0.398**	Supported
H2: Commitment to learning --> ICE	0.385**	Supported
H3: Learning style --> ICE	0.358**	Supported
H4: Absorptive capacity --> ICE	−0.012	Not supported
Type of industry	−0.124	−
Type of ownership	−0.033	−

**p<0.01

Discussions

While the corporate entrepreneurship literature is evolving and maturing over the past few decades, scholarly works on its international counterpart is still scarce (Lampe et al., 2020; Glinyanova et al., 2021). This study seek to address the void in the literature to provide empirical evidence on the role of entrepreneurship in international business. This study is among the few that specifically address an important yet overlooked question in the IB literature: can exporters become entrepreneurial by learning? This study specifically look into how entrepreneurship can be induced by organizational learning in an emerging market.

The results of this study shows that international corporate entrepreneurship is a pivot concept in explaining international firm performance. Exporters in China have a lot of opportunities to engage in international operations by exporting their products. However, they are also faced with high competition among themselves and also increasing demand from foreign buyers. The capacity to innovate is highly essential to ensure that they stay competitive.

It is found that organizational learning plays a crucial role when exporters engage in entrepreneurial activities. However, not all aspects of learning can enhance entrepreneurship: only commitment to learning and learning style can help foster entrepreneurial commitment, but absorptive capacity cannot. It appears that having an organizational culture that values learning is of vital importance and that learning has to be focused on double-loop learning. Given that the market is getting more dynamic, it is imperative that firms have to be able to interpret evolving changes in the market much faster than before. Armed with the determination to learn as well as a higher-level learning capability, exporters can more readily engage in innovation and undertake risky endeavors.

It is also found that absorptive capacity has no impact on ICE. Having an export team with prior experience and training in international operations does not help the firm to be more entrepreneurial. This seems to suggest that having experienced staff will not offer advantage to the firm and may even have a detrimental effect (as shown by the negative coefficient). Experience is actually a double-edged sword. On one hand, experience allows employees to quickly get hands-on with a new task if they have previously done this before or if they are trained to do so. However, they may also "rest on their laurels" in particularly if they enjoyed previous successes and this may actually deter innovativeness and risk taking.

Managerial Implications, Limitations and Future Research

For exporters, the results of this study can provide them with useful guideline in pursue of international expansion. In particular, being entrepreneurial is highly important. Exporting firms should always be ready to update its products/services to the ever-changing needs of the foreign markets. This is a risky endeavor so in order to minimize the risk, entrepreneurship has to be guided by an organizational-wide culture to continuously learn as well as a dedication to learn at higher level. In other words, an "open mind" is required to constantly on the outlook for opportunities and be ready to initiate changes when needed. Working experience and training appears to be eroding in their prominence as found in this study. In this hyper turbulent environment in which acquired knowledge becomes outdated rapidly, previous international experience or training may not be able to help in entrepreneurial undertakings. It is therefore not surprising to find new start-ups becoming immensely successful not because of previous experience but through entrepreneurial acts. Exporters therefore should be on the outlook for new opportunities, always engage in high level learning, updating their knowledge base continuously in order to stay competitive.

This study has a number of limitations which should be considered when interpreting the findings. While we argue that organizational learning is important in stimulating entrepreneurship in exporting firms, we have only included three aspects of organizational learning. Future studies will need to look at other aspects of learning to give a better picture of the impact of organizational learning on entrepreneurship. In addition, the sample includes a variety of firms including both service and manufacturing companies as well as of different ownership types and sizes. By focusing on a particular type of firm (e.g. SMEs), it can help avoid the confounding effects that may arise from differing firm characteristics and therefore better insights can be obtained. Obviously, the size of the sample can be enlarged to better represent exporters which are abundant in China. Moreover, given that this study is undertaken in the institutional context of an emerging market which is developing rapidly, it would be instrumental to include institutional factors into the research model to investigate the impact of these variables on international performance.

References

Alvarez, S.A., & Busenitz, L.W. (2001). The entrepreneurship of resource-based theory. *Journal of Management*, 27(6), 755–775.

Argyris, C., & Schön, D.A. (1978). *Organizational Learning: A Theory of Action Perspective*, Addison-Wesley: Reading, MA.

Augier, M., Shariq, S.Z., & Thanning Vendelø, M. (2001). Understanding context: its emergence, transformation and role in tacit knowledge sharing. *Journal of Knowledge Management*, 5(2), 125–137.

Autio, E., Sapienza, H.J., & Almeida, J.G. (2000). Effects of age at entry, knowledge intensity and limitability on international growth. *Academy of Management Journal*, 43, 909–24.

Baker, W.E., & Sinkula, J.M. (1999). The synergistic effect of market orientation and learning orientation on organizational performance. *Journal of the Academy of Marketing Science*, 27(4), 411–427.

Balabanis, G. I., & Katsikea, E.S. (2003). Being an entrepreneurial exporter: does it pay? *International Business Review*, 12, 233–52.

Barringer, B., & Bluedom, A.C. (1999). The relationship between corporate entrepreneurship and strategic management. *Strategic Management Journal*, 20(5), 421–444.

Bierwerth, M., Schwens, C., & Isidor, R. (2015). Corporate entrepreneurship and performance: A meta-analysis. *Small Business Economics*, 45, 255–278.

Brockmann, E.N., & Anthony, W.P. (1998). The influence of tacit knowledge and collective mind on strategic planning. *Journal of Management Issues*, 10(2), 204–222.

Brockmann, E. N., & Anthony, W. P. (2002). Tacit Knowledge and Strategic Decision Making. *Group & Organization Management*, 27(4), 436–455.

Burgelman, R.A. (1984). Designs of corporate entrepreneurship in established firm. *California Management Review*, 26, 154–166.

Calantone, R.J., Cavusgil, S. T., & Zhao, Y. (2002). Learning orientation, firm innovation capability and firm performance. *Industrial Marketing Management*, 31(6), 515.

Casson, M. (2000). *Economics of International Business*. Edward Elgar: Cheltenham.

Chandler, G., & Hanks, S.H. (1993). Measuring performance of emerging businesses: a validation study. *Journal of Business Venturing*, 8(5), 391–408.

Chaston, I., Badger, B. & Sadler-Smith, E. (2000). Organizational learning style and competencies: a comparative investigation of transactionally-oriented UK manufacturing firms. *European Journal of Marketing*, 34, 625–640.

Cohen, W.M. & Levinthal, D.A. (1990). Absorptive capacity: A new perspective on learning and innovation. *Administrative Science Quarterly*, 35, 1128–1152.

Covin, J.G., & Slevin, D.P. (1991). A conceptual model of entrepreneurship as firm behaviour. *Entrepreneurship Theory and Practice*, 16(1), 7–24.

Covin, J.G., & Slevin, D.P. (1989). Strategic management of small firms in hostile and benign environments. *Strategic Management Journal*, 10(1), 75–87.

Day, G.S. (1994). The capabilities of Market-Driven Organizations. *Journal of Marketing Management*, 10(8), 725–742.

Day, G.S. (1991). Learning About Markets. *Marketing Science Institute Report* Number 91–117, Marketing Science Institute: Cambridge, MA.

De Geus, A. (1988). Planning as Learning. *Harvard Business Review*, Mar/ Apr,74.

Dess, G., Ireland, R.D., Zahra, S.A. & Floyd, S. W. (2003). Emerging issues in corporate entrepreneurship. *Journal of Management*, 29(3), 351–378.

Dimitratos, P., & Plakoyiannaki, E. (2003). Theoretical foundations of an international entrepreneurial culture. *Journal of International Entrepreneurship*, 1, 187–215.

Drucker, P. (1954). *The Practice of Management*. Harper & Row: New York.

Evangelista, F. (2005). Qualitative insights into the international new venture creation process. *Journal of International Entrepreneurship*, 3(3), 179–198.

Fernández-Mesa, A. & Alegre, J. (2015). Entrepreneurial orientation and export intensity: Examining the interplay of organizational learning and innovation. *International Business Review*, 24(1), 148–156.

Fiol, C.M. & Lyles, M. (1985). Organizational Learning. *Academy of Management Review*, 10(4), 803–813.

Fornell, C., & Larcker, D. F. (1981). Evaluating structural equation models with unobservable variables and measurement error. *Journal of Marketing Research*, 18(1), 39–50.

Galer, G., & Van der Heijden, K. (1992). The learning organization: How planners create organizational learning. *Marketing Intelligence and Planning*, 10(6), 5–12.

Garvin, D.A. (1993). Building a learning organization. *Harvard Business Review*, 71 (July-August), 78–91.

Garvin, D.A., & Levesque, L.C. (2006). Meeting the challenge of corporate entrepreneurship. *Harvard Business Review*, Oct, 2–12.

Giamartino, G.A., McDougall, P.P., & Bird, B.J. (1993). International entrepreneurship: the sate of the field. *Entrepreneurship Theory and Practice*, 18(1), 37–42.

Glinyanova, M., Bouncken, R.B., Tiberius, V., & Cuenca Ballester Antonio, C. (2021). Five decades of corporate entrepreneurship research: Measuring and mapping the field. *International Entrepreneurship and Management Journal*, 17(4), 1731–1757.

Hamel, G., & Prahalad, C.K. (1994). *Competing for the Future*. Cambridge, Mass: Harvard Business School Press.

Hayes, J., & Allinson, C.W. (1998). Cognitive style and the theory and practice of individual and collective learning in organizations. *Human Relations*, 51(7), 847–871.

Hayes, J., & Allinson, C.W. (2005). Cognitive style and its relevance for management practice. *British journal of management*, 5(1), 53–71.

Huber, G.P. (1991). Organizational learning: the contributing processes and the literatures. *Organization Science*, 2, 81–115.

Hult, G.T.M., & Kandemir, D. (2003). Market orientation, learning orientation and innovativeness in the global marketplace: moderating roles of organizational memory and market turbulence. In: Subhash C. Jain (ed), *Handbook of Research in International Marketing*, 42–56, Edward Elgar Publishing.

Inkpen, A.C. (2000). Learning through joint ventures: A framework of knowledge acquisition. *Journal of Management Studies*, 37(7), 1019–1039.

Johanson, J., & Vahlne, J. (1990). The Mechanism of Internationalisation. *International Marketing Review*, 7(4), 11–24.

Johansson, J. and Vahlne. J. (1977), 'The Internationalization Process of the Firm: A Model of Knowledge Development and Increasing Foreign Commitments', *Journal of International Business Studies*, Spring: 23–32.

Kirzner, I.M. (1979). *Perception, Opportunity and Profit*. Chicago: University of Chicago Press.

Kogut, B., & Zander, U. (1992). Knowledge of the firm, combinative capabilities, and the replication of technology. *Organization Science*, 3, 383–397.

Korth, S. (2000). Single and double-loop learning: Exploring potential influence of cognitive style. *Organization Development Journal*, 18(3), 87–99.

Kropp, F., Lindsay, N.J., & Shoham, A. (2006). Entrepreneurial, market and learning orientations and international entrepreneurial business venture performance in South African firms. *International Marketing Review*, 23(5), 504–523.

Kuratko, D.F., Hornsby, J.S., & Hayton, J. (2015). Corporate entrepreneurship: the innovative challenge for a new global economic reality. *Small Business Economics*, 45, 245–253.

Kuratko, D.F., Montagno, R.V., & Hornsby, J.S. (1990). Developing an intrapreneurial assessment instrument for an effective corporate entrepreneurial environment. *Strategic Management Journal*, 11(1), 49–58.

Lampe, J., Kraft, P. S., & Bausch, A. (2020). Mapping the field of research on entrepreneurial organizations (1937–2016): A bibliometric analysis and research agenda. *Entrepreneurship Theory and Practice*, 44(4), 784–816.

Lant, T. K., & Mezias, S. J. (1992). An Organizational learning model of convergence and reorientation. *Organization Science*, 3(1), 47–71.

Lant, T.K. (1989). Aspiration level adaptation: an empirical study of three models. Working paper, New York University: New York.

Le, N.H., & Evangelista, F. (2007). Acquiring tacit and explicit marketing knowledge from foreign partners in IJVs. *Journal of Business Research*, 60(2), 1152–1165.

Lei, D., Hitt, M.A., & Bettis, R. (1996). Dynamic core competences through meta-learning and strategic context. *Journal of Management*, 22(4), 549-69.

Levitt, B., & March, J.G. (1988). Organizational learning. *Annual Review of Sociology*, 14, 319–340.

Liu, S.S., Luo, X.M., & Shi, Y.Z. (2002). Integrating customer orientation, corporate entrepreneurship, and learning orientation in organizations-in-transition: An empirical study. *International Journal of Research in Marketing*, 19, 369–382.

Liu, Y., Li, Y., & Xue, J. (2011). Ownership, strategic orientation and internationalization in emerging markets. *Journal of World Business*, 46(3), 381–393.

Lumpkin, G.T., & Dess, G.G. (1996). Clarifying the entrepreneurial orientation construct and linking it to performance. *Academy of Management Review*, 21(1), 35–172.

Matsuno, K., Mentzer, J.T., & Ozsomer, A. (2002). The effects of entrepreneurial proclivity and market orientation on business performance. *Journal of Marketing*, 66 (July), 18–32.

McDougall, P.P. (1989). International versus domestic entrepreneurship: new venture strategic behavior and industry structure. *Journal of Business Venturing*, 4, 398–400.

McDougall, P.P., & Oviatt, B.M. (2000). International entrepreneurship: the intersection of two research paths. *Academy of Management Journal*, 43, 902–908.

Miller, D. (1983). The correlates of entrepreneurship in three types of firms. *Management Science*, 29(7), 770–791.

Minniti, M., & Bygrave, W. (2001). A dynamic model of entrepreneurial learning. *Entrepreneurship Theory and Practice*, 25(3), 5–16.

Moorman, C. (1995). Organizational market information processes: Cultural antecedents and new product outcomes. *Journal of Marketing Research*, 32 (August), 318–335.

Morris, M., & G. Paul (1987). The relationship between entrepreneurship and marketing in established Firms. *Journal of Business Venturing*, 2(3), 247–259.

Naldi, L., Achtenhagen, L., & Davidsson, P. (2015). International corporate entrepreneurship among SMEs: A test of Stevenson's notion of entrepreneurial management. *Journal of Small Business Management*, 53(3), 780–800.

Naman, J.L., & Slevin, D.P. (1993). Entrepreneurship and the concept of fit: a model and empirical tests. *Strategic Management Journal*, 14(2), 137–53.

Nevis, E. C., DiBella, A. J., & Gould, J. M. (1995). Understanding organizations as learning systems. *MIT Sloan Management Review*, 36(2), 73–73.

Pellissier, J.M., & Van Buer, M.G. (1996). Entrepreneurial proclivity and the interpretation of subjective probability phrases. *Journal of Applied Business Research*, 12(4), 129–37.

Peterson, R., & Berger, D. (1972). Entrepreneurship in organizations. *Administrative Science Quarterly*, 16, 97–106.

Podsakoff, P.M., & Organ, D.W. (1986). Self-reports in organizational research: Problems and prospects. *Journal of Management*, 12, 531–544.

Podsakoff, P.M., MacKenzie, S.B., Lee, J.Y., & Podsakoff, N.P. (2003), 'Common method biases in behavioral research: A critical review of the literature and recommended remedies', *Journal of Applied Psychology* 88: 879–903.

Politis, D. (2005). The process of entrepreneurial learning: a conceptual framework. *Entrepreneurship Theory and Practice*, 29(4), 399–424.

Powell, W. W., Koput, K. W., & Smith-Doerr, L. (1996). Interorganizational Collaboration and the Focus of Innovation: Networks of Learning in Biotechnology. *Administrative Science Quarterly*, 41, 116–145.

Qiuqin, H., Wang, M., & Martínez-Fuentes, C. (2020). Impact of corporate entrepreneurial strategy on firm performance in China. *International Entrepreneurship and Management Journal*, 16(4), 1427–1444.

Rupčić, N. (2019). Entrepreneurial learning as individual and organizational learning. *The Learning Organization*, 26(6), 648–658.

Sackmann, S. A. (1991). *Cultural Knowledge in Organizations*. Newbury Park, CA: Sage.

Sambrook, S. and Roberts, C. (2005). Corporate entrepreneurship and organizational learning: a review of the literature and the development of a conceptual framework, *Strategic Change*, 14(3), 141–156.

Sathe, V. (1989). Fostering entrepreneurship in the large diversified firm. *Organizational Dynamics*, 18(1), 20–32.

Sawhney, M., Wolcott, R.C., & Arroniz, I. (2006). The 12 different ways for companies to innovate. *MIT Sloan Management Review*, 47(3), 75–81.

Schumpeter, J.A. (1939). *Business Cycle*. Vol. 1, New York: McGraw-Hill.

Schumpeter, J.A. (1950). *Capitalism Socialism and Democracy*. New York: Harper and Row:.

Senge, P. (1990). *The Fifth Discipline: The Art and Practice of the Learning Organization*. New York: Doubleday.

Shaw, R.B., & Perkins, D.N. (1991). Teaching organizations to learn. *Organization Development Journal*, 9(4), 1–12.

Simmonds, K., & Smith, H. (1968). The first export order: a marketing innovation. *British Journal of Marketing*, 2(Summer), 93–100.

Sinkula, J., Baker, W., & Noordewier, T. (1997). A framework for market-based organizational learning: linking values, knowledge, and behaviour. *Journal of the Academy of Marketing Science*, 25(4), 305–318.

Slater, S.F., & Narver, J.C. (1994). Does competitive environment moderate the market orientation-performance relationship? *Journal of Marketing*, 58(January), 46–55.

Slater, S. F., & Narver, J. C. (1995). Market orientation and the learning organization. *Journal of Marketing*, 59(July), 63–74.

Stevenson, H.H., & Jarillo, J.C. (1990) A paradigm of entrepreneurship: Entrepreneurial management. *Strategic Management Journal*, 11(summer special issue), 17–27.

Tobin, D. R. (1993). *Re-educating the Corporation: Foundations for the Learning Organization*. Essex Junction, VT: Oliver Wright.

Urbano, D., Turro, A., Wright, M., & Zahra, S. (2022). Corporate entrepreneurship: A systematic literature review and future research agenda. *Small Business Economics*.

van Doorn, S., Heyden, M.L.M., & Volberda, H.W. (2017). Enhancing entrepreneurial orientation in dynamic environments: The interplay between top management team advice-seeking and absorptive capacity. *Long Range Planning*, 50(2), 134–144.

Westhead, P., Wright, M., & Ucbasaran, D. (2001). The internationalization of new and small firms: A resource based view. *Journal of Business Venturing*, 16(4), 333–358.

Wolcott, R.C. & Lippitz, M.J. (2007). The four models of corporate entrepreneurship. *MIT Sloan Management Review*, 49(1), 75–82.

Wolff, J.A., Pett, T.L., & Ring, J.K. (2015). Small firm growth as a function of both learning orientation and entrepreneurial orientation. *International Journal of Entrepreneurial Behaviour & Research*, 21(5), 709–730.

World Bank (2020). Exports of goods and services. Retrieved: https://data.worldbank.org/indica
    tor/NE.EXP.GNFS.CD?locations=CN

Xinhua Net. (2015). Guangdong exports highest among seven biggest trade provinces. Available at
    http://www.gd.xinhuanet.com/newscenter/2015-08/09/c_1116190535.htm (accessed
    29 January 2016) (in Chinese).

Zaheer, S. (1995). Overcoming the liability of foreignness. *Academy of Management Journal*,
    38(2), 341–363.

Zahra, S. A. (1991). Prediction and financial outcomes of corporate entrepreneurship:
    An exploratory study. *Journal of Business Venturing*, 6(4), 259–285.

Zahra, S. A. (1993). A conceptual model of entrepreneurship as firm behavior: A critique and
    extension. *Entrepreneurship Theory and Practice*, 17(4), 5–22.

Zahra, S. A., & Garvis, D. (2000). International corporate entrepreneurship and firm performance:
    the moderating effect of international environmental hostility. *Journal of Business Venturing*,
    15, 469–92.

Zahra, S.A., & Covin, J.G. (1995). Contextual influences on the corporate entrepreneurship-
    performance relationship: A longitudinal analysis. *Journal of Business Venturing*, 10(1), 43–58.

Zahra, S.A., Jennings, D.F., & Kuratko, D.F. (1999). The antecedents and consequences of firm-level
    entrepreneurship: The state of the field. *Entrepreneurship Theory and Practice*, 24(2), 45–66.

Zehir, C., & Eren, M.S. (2007). Field research on impacts of some organizational factors on
    corporate entrepreneurship and business performance in the Turkish automotive industry.
    Journal of American Academy of Business, 10(2), 170–176.

Part IV: **Institutions and Innovation**

Gordon Redding and Chris Rowley

7 Societal Progress and Information: Impacts on Learning and Innovation for China and Beyond

Introduction

From 1980 for several decades China was host to one of the world's miracle surges of innovative economic growth. This achievement came about with the release of pent-up instincts to cooperate in seeking newly available advantages. It took place in a historical opportunity space that encouraged new commercial venturing, plus various fusions with external influences that had become welcome, notably the engagement of foreign industries and the bridging of cooperation with the regional ethic Chinese *diaspora* – themselves already globally integrated. The creation of 'special economic zones' were also often helpful in providing territorial separation for the experimental process supported by government. However, a broad question now increasingly asked about China is: can it restructure its economy from a low cost to a high value model? (Rowley & Oh, 2020). The answer to this is complex – with both macro and micro level elements in it – and learning and innovation to deal with greater complexity are critical components of it.

First, at the macro level, we can make several key empirical and theory-driven observations and points. All such societal growth phenomena work in surges and economic history teaches us that such surges usually last about 30 years. This is because the changes they bring with them cause a need to re-calibrate relations of many kinds with aspects of their context (Cardwell, 1972; Taylor, 2016; Janeway, 2018; McClellan & Dorn, 2015). The parts of the context usually affected include the politics, labour market, invited external companies, technology and new relations with the rest of the trading world. These new conditions and their implications for China became the subject of a large literature, a few examples of which are *inter alia*, Shambaugh (2013), Overholt (2018), Zheng and Huang (2018). A common theme in this literature is the need for all organisations – from the state down to the entrepreneur – to prepare for the growing complexity attached to a competitive modern economy.

In the middle of this potentially turbulent set of changing circumstances in China, the opportunities that became available themselves changed and so too did the pattern of entrepreneurial adaptation itself need to adjust. The resilience of business leaders was tested. This chapter examines evidence of that resilience at work within the quite distinct economic culture of China. How, in other words, did China's innovativeness remain refreshed?

https://doi.org/10.1515/9783110715002-007

Second, at the micro level, again we can make several key empirical and theory-driven observations and points. Innovation is intrinsically linked to learning and knowledge. These actually require careful nurturing and even managing. Therefore, the concepts, practices and management of learning and knowledge are important to discuss. Given this, we outline key, broad-brush points in these fields.

We begin with the wider context. This includes tracing globalisation and its underpinnings and China's situation in it and also historical development and the fourth industrial revolution, since that is a new global reality and it presents quite distinct challenges to any entrepreneur in almost any industry in any society.

'Time After Time': Globalisation's Changes

The word 'globalisation' is bandied about with abandon and used *ad nauseam* in a variety of fields. Being such an elastic concept, it is difficult to pin down its meaning and emergence. Taking a more historical view there is a variety of contenders for the first proto-globalists. For instance, the Romans constructed a network of robust, high quality roads across the then known ancient world. Alternatively, the 'Silk Roads' from the 1st century BC allowed luxury goods from China to travel to the other edge of the Eurasian continent – Rome (and trade reopened again in Marco Polo's late medieval times of the 13th to 14th century). For Hansen (2020) globalisation's 'big bang' began around 1000 when for the first time trade routes linked the entire globe and objects flowed across the world as the Vikings travelled extensively for trading, from Canada in the West to Turkey and the Ukraine in the East. Other contenders are Islamic merchants – as the religion spread, so did trade via the 'Spice Routes' of the 7th–15th centuries. By the 9th century Muslim traders dominated the Mediterranean and Indian Ocean and trade then stretched as far East as Indonesia and West as Spain. However, both spices and silk were luxury products traded in low volumes. Nevertheless, China's current Belt and Road initiative of trade between East and West already existed.

Another set of possibilities for globalisation's antecedence include the 'Age of Discovery' of the 15th to 18th centuries when explorers connected East and West. The Portuguese, Spanish and later Dutch and British discovered, subjugated and integrated new lands into their economies with global supply chains. However, this trade was narrow, siloed and lopsided, mostly with colonies and by exploitation, by creating a mercantilist and extractive and colonial economy, not a truly globalised one.

What the above indicates is that rather than seeing globalisation as somehow 'new' or even China-driven and dominated or a 'one off' or 'either-or' event, it is more useful to see it as a long historical process with ebbs and flows. These are led by technological forces in the fields of transport and communication, with decreasing barriers to the flow of trade, finance and people across geographic boundaries. Given this, we can see the earliest 'modern' globalisation from the 1870s, driven by the industrial revolution and its output and transport and communication innovations and huge global trade and free flow of capital and labour. This was ended by the first world word war in 1918 (Boyce, 2009). The subsequent 'de-globalisation' of the post-1920s lasted until the beginning of the new global economy of the 1950s–70s. One interesting difference from the 'golden age' of globalisation was that labour became less free flowing and barriers were erected. Driven in part by transport revolutions, such as containerization and air-freight modern globalisation developed from the 1980s and the level of globalisation increased rapidly between 1990 and 2007, but the Global Financial Crisis and the subsequent Great Recession impacted on it. Seeing globalisation as composed of different dimensions allows a more nuanced picture to emerge. Since 2015, globalisation continued to flatten out (KOF, 2018): economic globalisation stagnated, trade integration and cultural globalisation declined, while social globalisation increased only slightly, although financial globalisation continued to progress and global information flows continued to increase, while political globalisation increased the most (KOF, 2018).

Globalisation has in recent decades faced many challenges and issues (Diamond, 2018; O'Sullivan, 2019). These include the following. First, the slowdown in trade growth, as flows of overseas foreign direct investment have fallen, there are fewer and less profitable 'genuine' (ie earning at least 25% of revenues from abroad) multinational companies and the reshoring of supply chains as cost advantages decline. Second, less trade liberalisation since the Uruguay round (1995) and China joining the WTO (2001). Third, a shift to protectionism with a rejection by some countries of the principles of the global trading system.

We can bring this analysis up to date with the impact on globalisation of the Covid-19 pandemic and locate learning and innovation within it. Covid-19's economic consequences can be seen if we use four flow categories in globalisation. First, people, where free movement was increasingly constrained and then collapsed with some slow recovery. Second, money, which slowed. Third, goods/trade, which slowed although with some bounce back. Fourth, ideas/information, which did the opposite and surged due to the internet and the pandemic's intensification of the 'dash to digital'. For example, cross-border internet traffic jumped massively. So, the globalisation of information fuels e-commerce, making cross-border supply chains more flexible and boosting collaboration in areas such as vaccine research. The development of Covid-19 vaccines has demonstrated cross-border scientific collaboration with knowledge sharing, learning and innovation.

Globalisation – and within it China's evolution – can be further seen within theories used to explain economic development. First, Modernization Theory (Rostow, 1960), with its Weberian and Parsonsian antecedence, argues that all societies naturally progress through common stages of development from 'traditional' ('underdeveloped'), where 'mature' ('developed') ones started. Globalisation is part of a process of emulation and 'catch-up'. Second, Dependency Theory (Frank, 1966), argues that such linear progression is too simplistic and ethnocentric in a range of ways. Under-developed countries are not merely 'primitive' versions of developed countries but have their own unique features and structures and there are variations within them. It is biased in terms of its neo-liberal, Western assumptions and sees some societies as remaining dependent on others. In short, the political, social and cultural dimensions of development were neglected. The theory's Marxist aspects formed World Systems Theory (traced back to Karl Polanyi after 1918), seeing the global economy having a third category of countries, intermediate between the core and periphery, the semi-periphery (Wallerstein, 1974). Members of this group are industrialized, but with less sophisticated technology and less control of finances. Thus, globalisation is seen as in existence since the 16th century as before this global interdependence did not exist.

China's position in such models of globalisation can be seen if we compare the share of global gross domestic product (GDP) and manufacturing output over time with other leading economies – US, UK, Germany, France, Russia, Japan and the EU. An interesting picture emerges. Globalisation's 'Golden Age' (1870–1913) saw the decline of China and the rise of the US and Germany in these terms. In contrast, the post-1980 period saw the strong recovery of China and the steady relative decline of the US and Japan and steep fall in Europe. Clearly globalisation is not static or unilinear, therefore, theories have to be dynamic and flexible in order to help better account for changes, such as China's changing importance, centrality and position over time.

The Fourth Industrial Revolution

The revolutions of interest here are those normally seen as having been based on the emergence of new technologies, although these in turn have their own earlier history in the world of ideas (Rowley, 2021a). The first industrial revolution began in Britain in the late 1700s and then spread to Europe and the US. With a key take-off point around 1830 it was based on applying new sources of power, such as steam pressure controlled technically. This allowed railways, factory production and more reliable global shipping. The second technology-based revolution from around 1875 added steel making, electricity and heavy engineering. The third was the age of oil, the automobile and mass production from around 1908. The fourth, beginning around 1970 added information and communications technology and it continues to evolve, now unpredictably and in some senses overwhelmingly, as its main source of value shifts – from hardware to software.

In tracking these massive changes historically, Janeway (2018) suggests a subtle overall realignment as becoming inevitable. There has always been a 'three-player game' involving markets, speculators and the state. In this balancing act it has always been crucial for a state to be able to hold the totality together by claiming 'the over-riding power of a politically legitimate mission' (Janeway, 2018, p.294). As Mazzucato (2018) has explained, for most of the last 250 years the state in the high productivity economies has played an important supportive role in the workings of the economic system. It has done so mostly by public support of education, research and new technologies. Recent economic crises, together with the deeply invasive change now attached to technology itself and the destabilizing rise of inequality (Piketty, 2020), mean that the previous equilibrium is now being globally reconfigured. As Janeway (2018, p.294) concludes of the current revolution:

> the relationship between the IT sector and the state has been reversed. Dependent on state support of research and procurement through its growth to maturity, the IT sector has now fostered a full-fledged digital revolution, comparable in scale and scope to the consequences of the railroads and of electrification. And the resulting transformation of economic and social and political life now confronts the market and regulatory structures of the legacy economy and redefines the responsibilities of the state.

The Case of China

Having identified a global trend affecting relations between public and private sectors, it is necessary to add a note here on the special case of China as a state-controlled economy. As Zheng and Huang (2018) illustrate in detail, for most of its history China's economy has actually been in effect under state control. Rather than the state 'being in the market', the market has always 'been in the state'. Periods of exception have been rare.

The entrepreneur-driven surge we referred to at the start of this chapter has already been re-configured to return dominance to the state. Whether it will produce the combination of forces advocated by Mazzucato (2018) as making an 'entrepreneurial state' depends on how two challenges come to be met. One challenge was first identified by Alfred Marshall in 1890 when he defined economics as a 'branch of biology broadly defined'. This perspective, now in the form of theories of Complex Adaptive Systems (Holland, 1995; Beinhocker, 2007; West, 2017), suggests that complexity itself requires responses of greater empowerment within a society. These requirements inevitably carry political implications. The second challenge is related and it is the management of societal responsiveness, in some societies at huge scale, this latter factor being always for China both a liability brought by reduced control, and an asset brought by the coordination potential in planning, as with infrastructure.

A Wider Perspective

Societies that progress towards high income *per capita* have historically always done so by exercising a capacity to adjust themselves to fit with surrounding change; but to do so without abandoning their essential civilizational heritage. They each do this in their own way, hence the wide varieties of types of capitalism (Redding & Witt, 2007). A clear example is Japan. The wider pattern against which such a phenomenon is set has been described by Welzel (2013) as a movement along a continuum from a 'coercive' to an 'emancipative' state order.

With data resting on 60 years of the World Values research and reflected in other studies (e.g. Pinker, 2018; Acemoglu & Robinson, 2012; Mokyr, 2017; McCloskey, 2016; Fukuyama, 2014), a pattern is visible showing that prosperity is also associated with improvements in ethical standards that themselves encourage expression of the more benevolent human instincts of tolerance, sharing, cooperativeness, equality and a sense of justice. As Heilbroner (2015) has summarized the matter, the various successful forms of capitalism share the invention of a more benevolent form of domination than previously achieved. This does not preclude the emergence of new responses different from capitalism, if they can match its *per capita* productivity.

In relation to 'innovation', a study of societal progress by Sklair (1970) pointed to a crucial ingredient being not so much innovation itself, but the context of its being encouraged – thus 'innovativeness', a culture of enquiry, opportunity, encouragement. The literature cited above on societal progress testifies to the relevance of this 'atmosphere', while fully acknowledging that it can be expressed in many ways, depending on other influences within a single society that shape its interpretation.

A Micro Perspective

At a more micro level, innovation requires knowledge and its learning, requiring not only its creation, but also its efficient and successful transmission, storage, retrieval and application – both individually and organisationally. When it comes to learning by individuals and organisations, various relevant points can be made (Rowley & Poon, 2011b). The defining characteristics of learning is 'change', by either acquiring something new or modifying something that already exists and which must be long lasting to ensure learning has really occurred (Knowles, 1990). The focus of learning and its outcomes can include behaviour, skills, cognition or affect. We can further distinguish between 'single' and 'double' loop learning (Argyris & Schon, 1978), which relates to concepts of 'first' and 'second' order learning (Bateson, 1972). Such concepts are used in creating the 'learning organisation' (Senge, 2006) that encourages thinking within a set of principles and techniques. A learning organisation is where: people continually expand their capacity to create desired results; new expansive patterns of thinking are nurtured; collective aspiration is set free and people continually learn together. Thus, learning organisations 'facilitate the learning of all members and consciously transforms itself and its context' (Pedler et al., 1997, p.3). Sharing knowledge, experience and ideas becomes a habit in the learning organisation. In sum, all organisations learn, whether they consciously choose to or not – it is a fundamental requirement for sustained existence (Kim, 1993). The learning organisation is an instrument for connecting learning and knowledge management. An important core element of the learning organisation is the learning itself as it has the power to change people's perceptions, behaviours and mental methods (Marquardt, 2002). This, then, facilitates, encourages and maximises learning at individual and team level. The organisational context of encouragement is obviously crucial.

Organisational learning has been increasingly recognised as a critical factor for organisations (Garvin, 1993; Senge, 2006). The organisational learning process consists of the way people learn and work together to overcome changes, leading to better knowledge and performance. It involves experimentation, observation, analysis, willingness to examine both successes and failures and knowledge sharing (Watkins & Marsick, 1993). Organisational learning is also a human process involving individual willingness and social interaction to detect and correct errors for continuous improvement. Such a response is often termed 'critical thinking' – a notion that implies the encouragement of exchange, debate, and reciprocal learning (Rowley, 2021b).

According to Senge (2006) the practices of organisational learning can be expressed through five lifelong 'disciplines', a development path for acquiring certain skills and competencies. These are:

Personal Mastery: learning how to generate and sustain creative tension, it continually expands one's ability to create the results in life one truly seeks.

Mental Models: deeply ingrained assumptions, generalisations or even pictures and images that influence how people understand the world and how they take action.

Shared Vision: capacity to hold a shared vision of the future of 'what we seek to create'.

Team Learning: working together to gain insights from team members for innovative and co-ordinated action and engage in dialogue.

System Thinking: the discipline that integrates all the above and fuses them into a coherent body of theories and practices to enable collaborative actions by employees.

The organisational learning process can be seen in four ways. First, a socio-cultural perspective sees learning as embedded in local organisational cultures, norms and values that influence activities and it is a highly contested concept embedded within power relations within organisations (Contu & Wilmott, 2003). It also needs to transcend organisational divisions that might affect perspectives, as between production, marketing, finance, etc. Second, a cognitivist perspective sees individuals as processors of information. Given limited capacity to process information, individuals rely on representations of an outer environment to learn. Learning takes place through a negative feedback mechanism of adaptation through trial and error. Third, a behaviouralist perspective shows organisations become more efficient at doing something by repeatedly doing it. The problem is that organisations can become constrained by their own experience and consequently path-dependent. Fourth, a resource-based perspective suggests that knowledge and know-how are shared through activities where individuals tell and show how certain operations are carried out in the 'best way'. Organisational learning is then seen as being based on practical undertakings and standard operating procedures (Gherardi, 2000). These four perspectives are useful in that they encourage a fuller picture of organisational learning.

However, there remain debates about the organisational learning concept. For example, organisations are collections of people who each have a unique learning style and motivation to understand and plan for their interactions. Some studies show that there are at least 32 different learning styles (O'Connor, 1997) and hence the learning perspective is diverse.

Hong, Snell and Rowley (2017) provide a practically-oriented international overview of organisational learning. This discusses the role of organisational learning in addressing challenges posed by sources of surrounding uncertainty such as 'institutional voids' and government intervention some of which may entail managing the associated 'unlearning' processes. Importantly, it also discusses the role of imitation as a precursor to innovation and the concept of 'effectuation' sheds light on how organisations learn, such as about not only entry strategies and internationalisation, but also knowledge sharing and knowledge transfer. Critically, both practical illustrations and theoretical explanations of the need for adaptation of organisational learning systems and practices to fit distinctive contexts remains highly relevant (Hong, Snell & Rowley, 2017), not least for the distinct business environment of China. The concept of polychronic knowledge creation (Chin, Wang & Rowley, 2020) and innovation and new ways of creating and managing knowledge (Chin, Hu, Rowley & Wang, 2021) have been covered in relation to China and Asian business models.

Organisational learning is clearly a key aspect of knowledge management (Rowley & Poon, 2011a) and both are linked to innovation. Knowledge management facilitates the leveraging of intellectual capital or knowledge residing in people's minds and deals with the identification, acquisition and maintenance of organisational knowledge in two forms. First, 'explicit', implying it can be articulated so it can be documented, codified, easily classified, categorised, combined and distributed to others. Second, 'tacit', such as 'rules of thumb', intuition, tips and techniques, based on experience, accumulated over time and influenced by intangible factors such as personal beliefs, perspectives and values and difficult to express, communicate or share.

Knowledge management is important as the formalisation and means of access to experience, knowledge and expertise and in the process it can create new capabilities and also encourage innovation. Knowledge management is treated as the systematic, explicit and deliberate building, renewal and application of knowledge to maximise knowledge-related effectiveness and returns on knowledge assets. Knowledge management has three components: 'Creation', 'Sharing', 'Transfer'. Creation can be via four modes (Nonaka and Takeuchi, 1995): 'Socialisation'; 'Externalisation'; 'Combination'; 'Internalisation'. Sharing can occur in both formal and informal settings. Some (Collins, 2001) believe tacit knowledge is embedded in 'mental models', so knowledge sharing is not simply a matter of managing information, it is essentially a deeply social process, which must take into account human and social factors, as well as cultural issues (Clark & Rollo, 2001). Transfer involves distribution and dissemination between individuals or groups or in a 'community of practice'.

Innovation exists in a 'zone' of uncertainty and non-conformity, requiring inventiveness, ingenuity and imagination, as well as risk taking. Yet, this faces the inhibitors and constraints from the political and institutional environment, which in China's case has induced a focus on quantity, conformity, harmony and economic development. These limitations have been turbo-charged by the idea of 'involution' ('*nei juan*') resulting from the intense zero-sum competition of struggle over limited resources and resource competition – working harder and harder with no gain. This idea has gained ground and a growing following in China. (Yang, 2021). The term 'involution' was coined by anthropologist Goldenwater (1936) to describe a culture that cannot (or does not) adapt and or expand its economy but continues to develop only in the direction of internal complexity and inefficiency. Geertz (1963), another anthropologist, popularised the word in the 1960s in work describing unusual aspects of one rural economy in Java, Indonesia. This posited that, over the course of centuries, ever-increasing amounts of labour had poured in, but output remained constant and no innovation had occurred. This idea has enduring relevance and is an early exploration into World System Theory with its emphasis on agriculture and ideology.

Knowledge is a key component of innovation, but it may be constrained as organisations use too many structured approaches and tools to capture and disseminate knowledge and such formalities may stifle the development of innovation. This requires a focus on managing the supporting structures and climate that allow individuals to engage in interaction and communication, resulting in new knowledge and innovation (Murry & Blackman, 2006).

Innovation management is a still relatively unstructured field and is characterised by a confusing and contradictory diversity of approaches, prescriptions and practices. Innovation can be thought of as a process on a spectrum involving little change through to system transformation, including continuous improvement. Conceptualised as: 1) a 'process', innovation evaluation emphasises a series of stages and phases; 2) a 'product', companies emphasise the impacts and results derived from innovation activities and the performance of those activities can be measured according to the product. This activity view of innovation focuses on the technological and managerial implications (Liu & Tsai, 2007). 'After all, innovation comes from the acknowledgement of vicious circles and dead ends and the investment of positive and action-led approaches.' (Rowley & Poon, 2011a, p.121). So, there is a need '. . . for balance between what has to be made stable, structured and systematised and what has to be creative, dynamic and open' (Ibid.).

Adams et al (2007) propose that the three knowledge management areas important for innovation management are: idea generation; knowledge repository; information flow. First, the early stage of the innovation process is a fuzzy period, including idea generation, opportunity identification, data analysis, idea selection and concept development. Second, if knowledge is fundamental to innovation, it should be possible to measure the accumulated knowledge of the firm, in other words its knowledge repository. One aspect of innovation relates to the combinations of new and existing knowledge. Central to this perspective is the idea of 'absorptive capacity', the firm's ability to absorb and put to use new knowledge, recognise the value of new, external knowledge, assimilate it and apply it to commercial ends (Cohen & Levinthal, 1990). Firms with strong absorptive capabilities are more likely to acquire knowledge and learn effectively from outside. Higher levels of absorptive capacity are related to innovation (Chen, 2004). Third, innovation management involves information flow into and within the firm, as well as information gathering and networking in knowledge management.

What we have shown with this micro level analysis of management issues around knowledge, learning and innovation are the building blocks to the macro level studies. Without such understanding and practices in countries and their management and employment, then innovation can be seen as an obviously worthy ideal but only loosely implemented because of the high sophistication and sensitivity needed when ideas rather than objects are being shared. It can as a result fail to match all the state level assertion and even encouragement.

The Variety Within China

The past few decades of China's progress have seen much experiment and the evolving of three distinct forms of business system, themselves now in process of adjustment under the further strengthening of the state's role in the economy. Since their underlying dynamics remain influential it is useful to retain them as categories undergoing different forms of change. An analysis of cultural and institutional appropriateness in the face of rising complexity shows how they evolved into the three main responses. An explanation is provided in terms of Information Theory by Boisot, Child and Redding (2011). Here the three forms are seen as responses to the way in which a society deploys information. This key determinant of much else is seen as varying along two dimensions. First, the degree to which information becomes 'codified', in other words reduced to precise indicators that can be taken to represent a much wider set of features, an example being a firm's price-earnings ratio. Second, the degree to which such symbols of meaning come to be diffused through the society, as for instance in the financial pages of a newspaper. A person reading those pages would be able to comprehend quickly a great deal about a society's economy and the units inside it. Yet, the quality of this comprehension rests on the reliability of the symbols and in turn on the reliability of the inputs on which they are based, such as sets of audited accounts and in turn the honesty of owners in terms of public duty.

In any society with low levels of reliable codification and diffusion of information other responses to the need to reduce uncertainty remain in use. For China the most widespread of these occurs within the category named 'fiefs' based on interpersonal trust, as in the *guanxi* networks of the private sector (Nolan & Rowley, 2021) and managerial ties (Rowley & Oh, 2020). Another response is the co-opting of official support under reciprocal understandings, as in the regional networks that emerged from the dissolving of the earlier collectives of the Mao era and the emergence of 'local corporates'. The third sector, that of the state's 'bureaucracy', relies more on the rationality-based coordination needed to deal with scale and complexity and so it deals with information of a more structured kind, but without the degree of diffusion that would be normal in the 'market' condition, thus leading the system open to being seen as opaque (Piketty, 2020, p.660).

An indication of the negative effects on factor productivity in resource allocation and use in the context of innovation, when reliable information is weakly codified and diffused only partially, is available from studies of corporate governance (Tricker & Li, 2019, Braendle, Gasser & Noll, 2005) and of contrasts in productivity between the private and state sectors (Lardy, 2014). These indicate that the competitive rigours of a free-market system can be undermined when favouritism and collusion between interest groups interferes to prevent the prioritizing of rational efficiencies in the use of different forms of input: human, financial, and technical. This challenge is likely to be increased when a society exhibits low civic morality and zero-sum behaviour (Shambaugh, 2013). The counter-balancing 'bourgeois virtues' seen as being so significant in other socio-economic systems, are features not just of private entrepreneurs themselves, but of the society that observes and legitimises their behaviour (McCloskey, 2006).

The Case of Solar Panels

To counter-balance the impression of difficulty in the use of information to absorb rising industrial complexity, we now examine the case of China's solar panel industry, drawing on the account by Zhang and White (2016) of how entrepreneurs successfully overcame the liability of the industry's newness. A key to industrial success was ability to apply private entrepreneurship to the surrounding institutional framework in both government and finance. This entailed changing the rules for achieving access to key resources. These rules for success were perceived here as: (a) entrepreneurs' reputation management; (b) technology competence; and (c) finance. They were used to guide the achievement of one of China's greatest innovation successes.

This account of how that was done brings together perspectives from both Entrepreneurship Theory and Institutional Theory. It also supports findings from a wider literature that emphasizes the need for the group of key actors, such as entrepreneurs, to have enough autonomy to influence change in the surrounding system (Habermas, 1984; Sklair, 1970; McCloskey, 2016). However, behind this is a further qualification, identified earlier by Polanyi (1944) and reinforced recently by Mazzucato (2018) and Janeway (2018), which is that the contributing role of the state can be crucial. This contribution is not just that of financing strategic risk, but also in providing the *order* on which a market rests if it is to be efficient. As Polanyi (1944, p.144) pointed out of an earlier history: 'the road to the free market was opened and kept open by an enormous increase in continuous, centrally organized and controlled interventionism'. As to the research which becomes increasingly crucial – the funding of high-risk science, for instance in biotechnology, has in the developed world a 'public leader and a private laggard' (Mazzucato, 2018, p.73).

Global differences reflect varying interpretations of how these forces are balanced to be in accord with local ideals. It is, in other words, possible for a state to be either too interventionist, or alternatively not enough involved. The two forces, state control and entrepreneurial creativity, co-evolve, but across a spectrum of options affected by cultural understandings about authority and trust. Influential also is the level of an industry's dependence on science. The still unanswered question is whether there are, nevertheless, universals stemming from Complexity Theory that apply to all forms of such balancing.

The solar panel industry, as suggested, is one of China's great success stories. It grew when entrepreneur entrants into an existing industry created a new organisational form so as to access key resources. At the outset around 2000, the industry in China was dominated by four state-owned enterprises (SOEs), themselves reluctant to licence private upstarts. Yet, those SOEs at that time were operating below international standards of technology. Also, at that time the industry globally was in its early days and was dominated by large Japanese, American and German firms, all constrained by the high market price of any product and the limited supply of core components, such as high-grade silicon. A crucial turning point for the global market was the decision by Germany in 2004 to amend its feed-in tariff regulations. This change encouraged its import of foreign contributions. Another surge followed around 2007 when the global market began to take more serious note of 'green' issues. However, it was not until 2010 that the Chinese government officially cited solar panels as a strategic emerging industry deserving support.

The study by Zhang and White (2016) analyses the processes whereby the 10 largest Chinese solar panel firms evolved during that period of major change leading up to 2010. The account describes three forms of action that made their transition possible:

1. *Leveraging*: or building claims of relevant personal competence on the part of the owning founder; and claims of organisational relevance as agile firms in a market evolving rapidly in both technology and demand.
2. *Aligning*: or building claims to match the expectations of key supporting institutions, regulatory, normative and cognitive. This meant capacity building in fields such as reliability to support investor confidence, technical certification and accumulated expertise.
3. *Enacting*: or publicly promoting the achievements to date and the future vision, so as to change the perception of what is legitimate. For example, when one of the firms became the first private Chinese firm to list on the NYSE, making its owner the richest man in China in 2006.

Thus, 'the legitimacy of this new organisational form across a wide range of resource holders, including investors, technical and managerial talent, suppliers and government officials was solidified. It also attracted new entrepreneurs to the industry' (Zhang & White, 2016, p.611).

Seeing Opportunity

The role of prospecting for opportunity is not identified *per se* in the above account by Zhang and White (2016) but may be taken as implied in what they term *enacting*. Other classic accounts of entrepreneurship emphasize change, or what Schumpeter (1976) famously labelled 'creative destruction'. What firms might envisage as the key arenas of change where opportunities might be sought are conditioned by the various dependencies attached to any industry. For example, in the food industry the key is perceived product safety and quality. In fashion it is perceived reputation for current style. In the solar panels case the particular dependencies are three forms of access to: large-scale finance (and so political support); leading edge technology (also a political issue); and production facilities managed with high level expertise and efficiency. This trio of 'keys' became the agenda of the leading firms and their focussed attention on them led to major industrial change.

The private Chinese solar panel firm did not exist as an organisational form before 2002. Yet, within just five years it had displaced the big multinationals in the industry to become the dominant source of solar PV panels globally. A survey of the global market shares in 2020 (news.energysage.com 2021) showed that of the 10 leading firms in the industry, 7 were based in China with the majority being of private origin.

Processes of Learning and Innovation

In a study of societal processes using which a society meets Sklair's (1970) criterion of being 'innovative' and Eisenstadt's (1965) concept of 'transformative capacity', plus taking account of recent insights from Societal Evolutionary Theory, Redding and Drew (2016) identified a series of social mechanisms that could be classified as enabling or disabling for both the innovativeness and the cooperativeness needed, these two features being seen as reciprocally linked. The various societal formulae that evolve in different cultures to achieve this reciprocity are exercises in building what many theorists of societal progress agree to be the universally aspired-to achievement: a society's 'transformative capacity' (Eisenstadt, 1965) or its ability to reinvent itself.

These suggested universals are sevenfold:

1. The economic units should be capable of being scaled up, except in certain industries such as the craft or service sectors resting on individual skills.
2. Any scaling up should be achieved while retaining the psychological engagement and creative contributions of employees.
3. Such employees should work with an encouraged degree of autonomy.
4. Such employees should receive a fair (as perceived) reward for their contribution.
5. Information should flow through the society in a way that fosters both learning and collaboration.
6. The surrounding system of societal order and regulation should act protectively and predictably and express trusted and shared ideals.
7. There should be a fundamentally neutral position in decision-making about risk so that it rests on objective and rational calculation and evidence-based judgement. Innovativeness should be applied in conditions of competitive fairness.

The defining of these universals is only the beginning of a complex process of their achievement. The nature of their being applied is suggested by Lewin and Volberda (1999), who focus not so much on ideals such as those listed, but on the way in which those features become integrated into the workings of a society's business system. To understand this they focus on the enabling mechanisms that become unobservable socially embedded meta-capabilities internalized and assimilated by actors. As Fukuyama (2014) also suggests from his global studies of societal progress, there are levels of social mobilization and of societal capacity. As many other observers conclude, the challenge of moving upwards through the levels is one of maintaining social stability while encouraging empowerment (Pinker, 2018; Mokyr, 2017; McCloskey, 2016; Beinhocker, 2007; Cosmides & Tooby, 2006; Halpern, 2010; Heilbroner, 1985; Landes, 1998; North, 2005; Nowak, 2011; Ridley, 2011; Spence, 1990; Weber, 1930; Welzel, 2013)

The Larger Trends in China's Economy

Over the last three decades China's economy has become increasingly autarkic, in other words a self-sustaining political economy less and less dependent on other societies. As Blanchette (2021) describes it, this condition evolved in stages. The first phase followed accession to the World Trade Organisation (WTO) in 2001. In the five years to 2006 China's trade in goods and services grew from 38% of GDP to 65% as it took on the role of 'workshop of the world'. In those five years the compound average growth rate of imports was 26% as components were brought in to feed the new industries selling finished products to the world. The 2007 Asian Financial Crisis then burst that bubble and led to a follow-on policy for China of increasing import substitution. Value-added would be kept inside China. In the five years to 2020 the compound average growth rate of imports was 2%. Export growth has also slowed. Many multinational companies decided to take their profit inside China, but as Paterson (2021) notes there is operational risk in doing business in China for both cultural and policy reasons.

The leading current suppliers of components into China are Japan, Korea and Taiwan, but the global value chains in industries they contribute to are vulnerable to shifts of geography driven by changes in cost advantage and/or political changes. These are especially a threat as the world recovers from the Covid-19 global pandemic and many nations, from the US and UK to Germany and Japan, are encouraging the reshoring of production. Also, cross-border risk has prompted 'hedging': diversifying into multiple locations for value chains. Furthermore, the demographics of ageing are likely to affect the labour market adversely. If these occur alongside lower capital formation, they may prevent a big improvement in total factor productivity and so real growth.

Conclusion

China has moved from being a temporary 'backwater' among researchers of economic development and is returning again to capture attention. Part of this is the continued steady erosion of *Pax Americana* by the strengthening and deepening of Asian economic development, prowess and growing dominance and the shaping and emerging of a new *Pax Sinica*. With this for China's policy-makers has also come a shift in interest towards emerging issues, away from the early and initial ones, such as inward investment and joint ventures, now more to the growing need to remain competitive, such as via learning and innovation. However, China's politically-driven socialist market economy contains some belligerent tendencies and also some that hinder the necessary economic and product upgrading and value-added trajectory that requires learning and innovation in Chinese firms. Crossing that river, even one step at a time, is likely to present a great challenge.

References

Acemoglu, D., & Robinson, J. A (2012). *Why Nations Fail: The Origins of Power, Prosperity and Poverty*. London: Profile Book.

Adams, R., Bessant, J., & Phelps. R. (2007). Innovation management measurement: A review. *International Journal of Management Reviews*, 8(1), 21–47

Argyris, C. and Schon, D. (1978). *Organizational Learning: A Theory of Action Perspective*, Reading, MA: Addison-Wesley.

Bateson, G. (1972). *Steps to an Ecology of Mind*, New York: Ballantine.

Beinhocker, E. D (2007). *The Origins of Wealth: Evolution, complexity and the radical re-making of economics*. London: Random House.

Blanchette, J (2021). 'From "China Inc." to "CCP Inc.": a new paradigm for Chinese state capitalism'. Report Feb 21: Singapore, Hinrich Foundation.

Boisot, M., Child, J., & Redding, G (2011). Working the system: Toward a theory of cultural and institutional competence. *International Studies of Management and Organization*, 41(1), 63–96.

Boyce, R. (2009). *The Great Interwar Crisis and the Collapse of Globalisation*, London: Palgrave Macmillan.

Braendle U.C., Gasser, T., & Noll, J. (2005). Corporate governance in China: Is economic growth potential hindered by *guanxi? Business and Society Review*, 110(4), 389–405.

Cardwell, D.S.L. (1972). *Turning Points in Western Technology: A Study of Technology, Science and History*, New York: Science History Publications.

Chen, C.J. (2004). The effect of knowledge attribute, alliance characteristics and absorptive capacity on knowledge transfer performance. *R&D Management*, 34(3), 311–21.

Chin, T., Wang, S., & Rowley, C. (2020). Polychronic knowledge creation in cross-border business models: A sea-like heuristic metaphor. *Journal of Knowledge Management*, 25(1), 1–22.

Chin, T., Hu, Q., Rowley, C., & Wang, S. (2021). Business model in the Asia-Pacific: Dynamic balancing of multiple cultures, innovation and value creation. *Asia Pacific Business Review*, 27(3), 331–341.

Clarke, T., & Rollo, C. (2001). Corporate initiative in knowledge management. *Education and Training*, 43(4/5), 206–14.

Cohen, W.M., & Levinthal, D.A. (1990). Absorptive capacity: A new perspective on learning and innovation. *Administrative Science Quarterly*, 35(1), 128–52.

Collins, H.M (2001). Tacit knowledge, trust and the Q of sapphire. *Social Studies of Science*, 31(1), 71–85.

Contu, A., & Wilmott H. (2003). Re-embedding situatedness: The importance of power relations in learning theory. *Organization Science*, 14(3), 283–96.

Cosmides, L. & Tooby, J. (2006). Origins of domain specificity: the evolution of functional organization. in: J.L. Bermudez (ed.), *Philosophy of Psychology*, London: Routledge.

Diamond, P. (Ed) (2018). *The Crisis of Globalisation: Democracy, Capitalism and Inequality in the 21st Century*. I.B.Tauris. London: Bloomsbury.

Eisenstadt, S. N. (1965). Transformation of social, political, and cultural orders in modernisation. *American Sociological Review*, 30(5), 659–673.

Frank, A.G. (1966). The development of underdevelopment. *Monthly Review*, 18(4).

Fukuyama, F (2014). *Political Order and Political Decay*. London, Profile Books.

Garvin, D. (1993). Building a learning organisation. *Harvard Business Review*, 71(4), 78–91.

Geertz, H. (1963). *Agricultural Involution: The Processes of Ecological Change in Indonesia*, Berkley, CA: University of California Press.

Gherardi, S. (2000). Practice-based theorizing on learning and knowing in organizations. *Organization*, 7(2), 211–23

Goldenwater, A. (1936). Loose ends in a theory on the individual and involution in primitive society; in Lowie, R.H. *Essays in Anthropology Presented to A.L. Kroeber*, Berkley, University of California Press.

Habermas, J. (1984). *The Theory of Communicative Action: Reason and the Rationalisation of Society*, London: Heinemann.

Halpern, D (2010). *The Hidden Wealth of Nations*. Cambridge: Polity Press.

Hansen, V. (2020). *The Year 1000: When Explorers Connected the World and Globalization Began*. New York: Scribner.

Heilbroner R.L. (1985). *The Nature and Logic of Capitalism*, New York, Norton.

Hampden-Turner, C., Abelin, R., & Rowley, C. (2020). How might culture hinder or help innovation? The culture clash between the formality and the informality needed to innovate. *Culture and Empathy*, 3(1–2), 3–22.

Holland, J. H (1995). *Hidden Order: How adaptation Builds Complexity*. New York: Basic Books.

Hong, J., Snell, R., & Rowley, C. (2017). *Organisational Learning in Asia: Issues and Challenges*, Oxford: Elsevier.

Janeway, W.H. (2018). *Doing Capitalism in the Innovation Economy*, Cambridge: Cambridge University Press.

Kim, D. (1993). The link between individual and organizational learning. *Sloan Management Review*, 35(1), 37–50.

Knowles, M.S. (1990). *The Adult Learner: A Neglected Species*, Houston, TX: Gulf.

KOF Swiss Economic Outlook (2018). KOF globalisation index: Globalisation lull Continues. *Press Release*, 27 December.

Landes, D (1998). *The Wealth and Poverty of Nations: Why Some are so Rich and Some so Poor*. New York: Norton.

Lardy, N. R. (2014) *Markets over Mao: The Rise of Private Business in China*, Washington DC, Peterson Institute for International Economics.

Lewin, A. Y., & Volberda, H. W (1999). Prolegomena on coevolution: A framework for research on strategy and new organizational forms. *Organization Science*, 10(5), 519–34.

Liu, P.L., & Tsai, C.H. (2007). The influence of innovation management on new product development performance in Taiwan's Hi-Tech industries. *Research Journal of Business Management*, 1(1), 20–9.

Marquardt, M. (2002). *Building the Learning Organisation: Mastering the Five Elements for Corporate Learning*, New York: Davies-Black Publishing.

Mazzucato, M. (2018). *The Value of Everything: Making and Taking in the Global Economy*. London: Allen Lane.

McClellan, J.E., & Dorn, H. (2015). *Science and Technology in World History*. Baltimore: Johns Hopkins University Press.

McCloskey, D. N (2006). The Bourgeois Virtues: *Ethics for an age of Commerce*. Chicago: University of Chicago Press.

McCloskey D.N. (2016). *Bourgeois Equality: How Ideas, not Capital or Institutions Enriched the World*. Chicago: University of Chicago Press.

Mokyr, J (2017). *A Culture of Growth: The Origins of the Modern Economy*. Princeton NJ: Princeton University Press.

Murray, P., & Blackman, D. (2006). Managing innovation through social architecture, learning and competencies: A new conceptual approach. *Knowledge and Process Management*, 13(3), 132–43

O'Connor, T. (1997). Using learning style to adapt technology for higher education. *Centre for Teaching and Learning*, Indiana State University.

O'Sullivan, M. (2019). *The Levelling: What's Next after Globalisation*, New York: Public Affairs.

Nolan, J., & Rowley, C. (eds) (2021). *Guanxi in Contemporary Chinese Business*. London: Routledge.

Nonaka, I., & Takeuchi, K. (1995). *The Knowledge Creating Company*. New York: Oxford University Press.

North, D. C (2005). *Understanding the Process of Economic Change*. Princeton: Princeton University Press.

Nowak, M. A (2011). *Super Co-operators: Altruism, Evolution, and Why We Need Each Other to Succeed*. New York: Free Press

Overholt, W.H. (2018). *China's Crisis of Success*. Cambridge, Cambridge University Press.

Paterson S. (2021). How much will China grow as an export market? Research Report 23 March 2021, Singapore, Hinrich Foundation.

Pedler, M., Burgoync, J., & Boydell, T. (1997). *The Learning Company: A Strategy for Sustainable Development*. London: McGraw Hill.

Piketty T. (2020). *Capital and Ideology*. Cambridge, MA: Belknap Press Harvard.

Pinker, S. (2018). *Enlightenment Now: The Case for Reason, Science, Humanism, and Progress*. London: Allen Lane.

Polanyi, K (1944). *The Great Transformation*. Boston: Beacon Press.

Redding, G., & Drew, A. (2016). Dealing with the complexity of causes of societal innovativeness: social enabling and disabling mechanisms and the case of China. *Journal of Interdisciplinary Economics*, 28(2), 107–136.

Redding, G., & Witt, M. A (2007) *The Future of Chinese Capitalism: Choices and Chances*. Oxford: Oxford University Press.

Ridley, M (2011). *The Rational Optimist*. London: Fourth Estate.

Rostow, W.W. (1960). *Stages of Economic Growth: A Non-Communist Manifesto*. Cambridge: Cambridge University Press.

Rowley, C. (2020), 'Perspectives on work, employment and management: Asia, comparisons and convergence', *International Studies in Management and Organisation*, 50(4), 303–316.

Rowley, C. (2021a). Managing people & technological change in context. In P. Kumar, A. Agrawal and P. Budhwar (eds) *Human & Technological Resource Management: New Insights into Revolution 4.0*, Emerald: xvii–xxii.

Rowley, C. (2021b). Culture and uncertainty: Meanings, reasons and results. *Culture and Empathy*, 4(2): 88–99.

Rowley, C., & Poon, I. (2011a). Knowledge management. In C. Rowley and K. Jackson (eds) *Human Resource Management: The Key Concepts*, London, Routledge, pp. 118–121.

Rowley, C., & Poon, I. (2011b). 'Organisational learning' in C. Rowley and K. Jackson (eds) *Human Resource Management: The Key Concepts*, London, Routledge, pp. 160–165.

Rowley, C., & Oh, I. (2020). The enigma of Chinese business: Understanding corporate performance through managerial ties. *Asia Pacific Business Review*, 26(5), 529–536.

Schumpeter, J. A (1976). *Capitalism, Socialism and Democracy*. London: Routledge.

Senge, P. (2006). *The Fifth Discipline: The Art and Practice of the Learning Organization*. London: Doubleday.

Shambaugh, D. (2013) *China Goes Global: The Partial Power*. Oxford: Oxford University Press.

Sklair, L. (1970). *The Sociology of Progress*. Abingdon: Routledge.

Spence, J. D. (1990). *The Search for Modern China*. London: Hutchinson.

Taylor M. Z. (2016). *The Politics of Innovation: Why Some Countries are Better than Others at Science and Technology*, Oxford, Oxford University Press.

Tricker R., & Li, G. (2019). *Understanding Corporate Governance in China*. Hong Kong: Hong Kong University Press.

Wallerstein, I. (1974). *The Modern World System I: Capitalist Agriculture and the Origins of the European World-Economy in the Sixteenth Century*. New York: Academic Press.

Watkins, K.E., & Marsick, V.J. (1993). *Sculpting the Learning Organisation*. San Francisco, CA: Jossey-Bass.

Weber, M (1930). *The Protestant Ethic and the Spirit of Capitalism*. London: Unwin.

Welzel, C (2013) *Freedom Rising*. Cambridge, Cambridge University Press.

West, G. (2017) *Scale: The Universal Laws of Life and Death in Organizisms, Cities, and Companies*. London. Weidenfeld and Nicolson.

World Bank/China State Council (2013). *China 2030: Building a Modern, Harmonious, and Creative Society*. Washington DC, World Bank.

Yang, Y. (2021). "Obedience and Fear": The Brutal Working Conditions behind China's Tech Boom. *Financial Times*, 9 June.

Zhang, W., & White, S. (2016). Overcoming the liability of newness: The emergence of China's private solar panel photovoltaic firms. *Research Policy*, 45, 604–17.

Zheng, Y., & Huang, Y. (2018). *Market in State: The Political Economy of Domination in China*. Cambridge: Cambridge University Press.

Javier Calero Cuervo and Vanessa Madalena Crestejo

8 The Path of Innovation in Macau's Integrated Resort Companies: Past Assessment for Future Scenarios

Introduction

Given the Chinese' penchant for games of chance, it is not surprising that Macau's privileged monopoly position in being the sole provider of legalized gambling in China, has led to being ranked 1st in the contribution of the "tourism and travel" sector to the region's economy. It was reported by the *World Travel & Tourism Council* that the sector's direct and indirect contribution to Macau's economy was 82.5% in 2008 – the highest amongst 176 tourism destinations (World Travel & Tourism Council, 2008). This high dependence of the economy from one market and on one sector has created greater vulnerabilities for Macau's Integrated Resorts and economy, such as from the disruption caused by the corona virus since late 2019.

As early as 2006, the concern of one sector's critical contribution to Macau's economy was given focus in the policy address of 16 November 2006 by the Chief Executive of the Macau SAR Government, Edmund Hau Wah Ho, who noted: "Recently, the rapid development of the local gaming industry has become a concern to more and more people, reflecting the maturity and far-sightedness of the whole community. As we implement adequate diversification of the economy, it is important to determine how to review and manage the gaming industry's development." (MSAR, 2007). The call for diversifying the structure of Macau's economy has been in every year's policy address but to no avail.

To assess diversification, various statistical measures on diversification of Macau's economy was undertaken by the Statistics and Census Service of the MSAR Government with annual data starting only in 2016 as reported in the *Analysis Report of Statistical Indicator System for Moderate Economic Diversification of Macao 2018* (DSEC, 2019). To put things in perspective, in this trilingual report on diversification (in Chinese, Portuguese, and English), the word "innovation" or its related words "innovative", "innovations", "innovate", and "innovatively", did not appear at all in the English translated part of the report; whereas diversification (or diversify) appeared 128 times.

The current Chief Executive of Macau SAR Government, Mr. Ho Iat Seng spoke candidly in his first policy speech of 20 April 2020: "Since Macao's reunification with the motherland, although the economy has experienced a stage of relatively rapid growth, the dominance of the gaming industry has not been reduced, but intensified." He also observed that "The past MSAR Governments have made efforts to promote economic diversity over the years but achieved little progress." (MSAR, 2020).

https://doi.org/10.1515/9783110715002-008

Yet, in a recent study, Liu and Lin (2022, p. 4) continue to propose that "economic diversification is the only way for Macau to achieve sustainable development. The path to promoting Macau's economy's moderately diversified and sustainable development lies in internal industrial pluralism and external regional deepening."

The efforts toward the importance of innovation for Macau's economic development was given greater prominence in the policy address of 15 November 2018 (MSAR, 2019) and significantly so, more recently on December 2021 with "The Second Five-Year Plan for Economic and Social Development of the Macao Special Administrative Region (2021–2025) of the Government of MSAR" (MSAR, 2021).

What could be the reasons for not being able to diversify Macau's economy? This study posits that one reason is because the efforts toward diversifying the economic structure of Macau was not sufficiently underpinned by innovations, in general; and to the effective applications of innovations to the integrated resorts and their businesses and operations. As history has shown in various ways and situations, necessity is the mother of inventions and innovations; whereas a privileged protected monopoly situation is the father of vulnerabilities and unsustainable advantages.

The next section discusses significant literature on diversification and innovation, in general; and its relevance toward the hospitality and tourism sectors. This is followed by an assessment of diversification and innovation in the past twenty years based on interviews of key informants from the industry, and secondary published materials. Finally, as Macau's hospitality and tourism sector is currently at a watershed period with the renewal of the gaming concessions, future driving forces and questions are put forward for the Integrated Resort companies.

Foundational Concepts: Diversification, Innovation, and Learning

Not being clear about the difference and relationship between diversification with innovation can create confusion and misdirected efforts by policy makers and managers. From a firm-level perspective, Diversification is "the process of firms expanding their operations by entering new businesses." (Dess, et al., 2016, p. 183); whereas Innovation is "the use of new knowledge to transform organizational processes or create commercially viable products and services." (Dess, et al., 2016, p. 382).

Diversification may not necessarily lead to the profitable operations of the new businesses. Likewise, efforts toward innovating will need to incorporate learning from the past. *Innovating* without learning from the past is to waste resources; whereas *learning* from the past without innovating is to neglect opportunities for a successful future. Vulnerable companies are those that do not invest much on both learning and innovation and are subject to greater risks when their privileged protected position are much reduced. Sustainable companies, in contrast, have wisely invested in innovating for their future while constantly learning from their past experiences (see Figure 8.1).

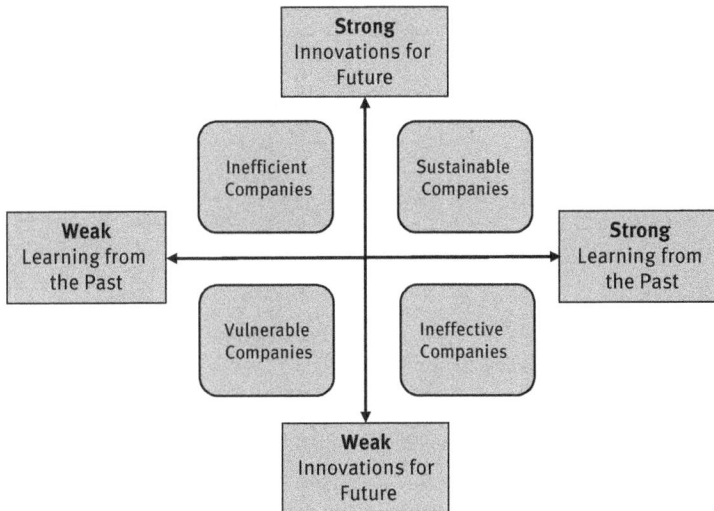

Figure 8.1: Levels of innovation and learning in four archetype companies.

There are various aspects of innovation that companies focus on or successfully combine, such as in the areas of business model, customer, product, technology, process, and organization (Greeven & Yip, 2021). An important and related concept is innovativeness. Innovativeness happens when managers have an attitude and willingness of nurturing and supporting an environment that allows for experimentation and creativity under an overarching entrepreneurial strategy (Dess, et al., 2016). These six aspects of innovation will be assessed with reference to the Integrated Resort companies in Macau (Greeven & Yip, 2021).

Based on an extensive, scientifically carefully selected process, and historical review of the literature, 209 academic journal articles and 145 trade journal articles on innovation in the hospitality and tourism sectors from 2010 to 2020, Cao, Shi and Bai (2022), identified the most prominent keywords that were related to the research done on innovation. These keywords are presented in Table 8.1 in the year they first appeared in these journal articles from 2010 to 2020.

As innovations become readily available, integrated resorts need to see to it that they are usefully and feasibly applied for the satisfaction of their key segmented customers and target markets. The question to ask is "Which innovations are necessary for sustaining our businesses?" Digital technologies such as Casino Management Systems, Electronic Table Games, Table Yield Management systems, RFID-embedded gambling chips, Angel Eye technology, and Thermal Security Cameras have been

Table 8.1: Innovation related keywords: Year first published in Journals on *Hospitality and Tourism*, 2010–2020.

Year	Academic Journal Articles	Trade Journal Articles
2010	ICT (Information Communications Technology)	APP, Menu, Mobile
2011	Hotel Service Innovation	Hotel booking, In-room facilities
2012	Sustainable	*No new keywords identified*
2013	E-WOM (Electronic Word of Mouth)	Restaurant, Employee, OTA (Online Travel Agency)
2014	Self-service Technology	Devices and Systems, Wearable mobile devices, Green Tech, Social Media
2015	Experience, Social Media	*No new keywords identified*
2016	UGC (User Generated Content), RFID (Radio Frequency Identification), Service Innovation	VR (Virtual Reality)
2017	Airbnb	AI (Artificial Intelligence), Digital, Virtual Assistant, Airbnb
2018	Smart Tourism	Robot
2019	OTA, VR, Open Innovation, AR (Augmented Reality)	Hotel electronic lock
2020	Digital, Employee, Robot	*No new keywords identified*

Source: Cao, Shi, and Bai (2022, Figure 5, p. 3804), Note: Table 8.1 is derived from an ocular inspection of Figure 5.

successfully applied in various integrated resorts in Macau for their gaming businesses and these have benefitted and provided value to both their customers and to the company (Liu, Dong, & Zhu, 2021). From the experience of the customers, Liu, Dong, and Zhu (2021, pp. 1687–1688) observed that these digital innovations "increases the diversity of product offerings, reduces waiting ties, ensures the transparency and fairness of the games and provides a safe and secure environment." From the integrated resort company's perspective, these digital innovations have enhanced the surveillance, security and speed of the casino's table games and slot machines; while providing management with better information about customers and guests through big data analysis for better business strategies (Liu, Dong, & Zhu, 2021).

Likewise, an equally and perhaps a much more important question for Macau is, "Are these innovations for the hospitality and tourism sectors being developed organically from Macau and are these commercially viable intellectual property rights being protected and purchased by various customers from around the world?"

Regarding intellectual property and tourism, the World Intellectual Property Organization (WIPO) explains how trademarks, copyrights, patents, geographical indications, and designs – Intellectual Properties – may be helpful for: "creating a distinctive identity in the market; protecting your competitive advantage; promoting national culture and heritage; or adding a revenue stream that would otherwise have not been possible." (WIPO, 2021). A comprehensive report with case studies of effective application of intellectual properties in tourism from various tourism destinations were provided in the WIPO website on "Boosting Tourism Development through Intellectual Property" jointly published by the World Intellectual Property Organization (WIPO) based in Geneva and the World Tourism Organization (UNWTO) based in Madrid on 2021 (WIPO and UNWTO, 2021). Given the important role of tourism for the economy of Macau, it was striking that there was no mention in the report of cases of intellectual properties developed in Macau for the tourism, leisure, and hospitality industry.

The Director General of WIPO, Daren Tang noted in his speech during the opening ceremony of the Guangdong-Hong Kong-Macao Greater Bay Area Intellectual Property Trade Expo 2020 held in Guangzhou on November 13, 2020 and published in the WIPO website that "according to WIPO's 2020 Global Innovation Index (GII), the Shenzhen-Hong Kong-Guangzhou cluster ranks second amongst the global top 100 science and technology clusters. The Hong Kong special administrative region as an economy ranks 11th in the GII. (. . .) Mr. Tang pointed out that intellectual property is one indicator of the innovative capacity and the Greater Bay Area shows dynamism and diversity in this area. Hong Kong SAR has a century-old IP infrastructure and is still building up its IP financing ecosystem. Guangdong boost a patent examination cooperation center, a specialized IP court, and an IP training platform. Macao shows great potential in IP and tourism-related projects." (WIPO, 2020). Indeed, Macau is very much in its nascent stage when it comes to innovation. Macau was not even ranked amongst the 132 economies in the Global Innovation Index 2021 (WIPO, 2021). Nonetheless, an assessment about innovation in the industry from key informants (KIs) with extensive experience in the operations of Integrated Resorts in Macau will be presented and analyzed.

Methods

A qualitative approach to the research was undertaken in order to have rich indepth insights of the phenomenon of innovation in Macau's Integrated Resorts. Interviews of managers who have ten to fifteen years of work experience at the Integrated Resorts in Macau was undertaken. There were four of the six managers who accepted the invitation to be interviewed. Two of whom were locals and two were foreigners and they worked for four of the six integrated resorts in Macau. The interview commenced with the same question asked of the managers, "How has innovation been implemented in Macau's six Integrated Resorts in the past twenty years?"

Follow up questions were asked as the interview progressed and the duration of the interviews ranged from one to two hours. All the interviews occurred in June 2022, of which two had to be conducted via Zoom due to the required new measures to prevent the spread of the virus in Macau. Three of the managers agreed to have the interview recorded, while one of the face-to-face interviews preferred the traditional approach of note-taking. As the content of the responses in this exploratory study is more relevant than who is saying what, there will be no specific attribution of the cited text from the interviews to any specific manager's profile as they all chose to remain anonymous.

Findings from Interviews of IR Managers – The Key Informants (KI)

In this section, the interviews will be related and analyzed with reference to the six paths of innovation of business model, customer, product, technology, process, and organization (Greeven and Yip, 2021) and with the specific literature concerning innovation in the businesses and operations of integrated resorts.

Business Model Innovation: Manifestations of Entrepreneurship

Tourism and hospitality companies in Macau need to innovate beyond their limited business diversification efforts as was pointed out by Greenwood & Dwyer (2017, p. 585): "To date, much of what is regarded as 'diversification' consists of attempts to expand the fringes of the market or to provide new attractions along the margins, rather than creating new revenue streams to challenge gambling." Indeed, it is important that diversification be supported through innovations that are commercially viable (Dess et al., 2016).

Business model innovations or what Hjalager (2010) categorized as "Managerial Innovations" would include a number of manifestations. One would be in the new ways of motivating staff in order to be loyal to the enterprise thereby minimizing employee turnover. This is especially a concern for the SME enterprises providing hospitality services. A second manifestation found in large companies, which would include some large Integrated Resorts amongst them, is the critical concern of controlling costs without jeopardizing the brand of the company. Another would be maintaining the flexibility of the tourism, leisure, and hospitality companies to be effective in meeting the various changing goals of the organization because of a market that has become dynamic and complex.

For most innovations to be effective, the quality of the people in the organization has to be the ones to realize its potencies. Integrated resort companies will need to strike the right balance and not jeopardize their business innovation efforts due to a failure to attract high-quality employees and suffer the consequences of high employee turnover that will eventually increase costs. As was rightly put by Brien, Vidwans and Dutt (2022, p. 6): "While low pay could be considered a positive productivity factor (low input costs), it may not attract high-quality employees who could provide innovation in services and thus increase productivity." On the decision of the recruitment of foreign employees vis-à-vis locals, one manager (KI3) opined that the IRs should *Ensure that Macau people are globally competitive in terms of skills, attitude and in delivering a quality of service and experience that reflects the Macau brand. This ensures that over time that diversifying and developing and creating more activities are Macau originated, particularly in retail, food and beverage & entertainment* would benefit the IR companies.

On the importance of controlling costs, one manager interviewed (KI1) shared their company's approach on how they identify areas for costs reduction:

> *Basically to look at the whole business eco-system from the customer and we move back to the supplier (i.e. the consultant – us). And then, we have a business focus – we have to define where it starts and where it ends. Then people look at what we call in lean [management] as wasteful activities.*

And for large Integrated Resorts, it is necessary to be efficient and reduce the overhead that comes with their extensive operations. On being efficient, the manager (KI2) noted:

> *Efficiency we do. Being a huge IR – there is a lot of overhead (air con, boiler, lights, etc.). We have a huge team but we also have the technology implemented to look over to manage and to have a centralized control to do savings (e.g. electricity, water, air-conditioning) through automation. How much does it costs for electricity or water and then try to manage based on the weather and the number of people we have in the property to make automatic adjustments: make the water hotter or colder; or to have less lights or more lights is all automated. That's the technology we need to implement in a large IR.*

Being too focused on the output while neglecting the process and learning involved in the various innovations developed in companies would be risky. As noted by Buitztendijk, van Heiningen, and Duineveld (2021, p. 9): "A one-sided focus on output, thus, risks failing to grasp these other important functions of innovation. Process-based indicators, such as indicators that capture the ability of an organisation to reflect on its own practices and learn from its successes and failures, are equally important." Furthermore, business innovations are manifestations of entrepreneurship and on the organic knowledge management process found in companies which are needed for developing smart tourism (Williams, Rodriguez, & Makkonen, 2020). A manager interviewed (KI1) corroborates this by sharing that

> *The key point for me as a Learning & Development manager was to create a sense of service – internal that would come from them. So, you can see some of the ideas of emergent innovation that is from the people who work in the company – organic development of innovation. In that*

*sense there would be attachment to the brand that would come from the employees – the busi-
ness ownership that was created – and catered for the company.*

Based on their research on organizational flexibility (OF), Anning-Dorson and Nyamekye (2020, p. 618) suggests that ". . . innovativeness must seek to create a form
of organizational adaptation to be transmitted into competitive advantage. The
findings suggest that to adequately achieve this, hospitality firms should seek to
increase their OF through innovativeness and to be able to rapidly identify market
trends, adjust internal operations, and respond quickly to new market demand."
However it might be good to point out that organizational flexibility needs as well
an effective learning process and knowledge management system that takes place
to effect business innovations. As aptly described by Wu (2020, p. 911): "Firms with
strong absorptive capacity are able to absorb or acquire new generated knowledge,
incorporating it with firms' prior knowledge and using it in the innovation process."
Unfortunately, one of the interviewed managers (KI3) observed that market research in Macau was hampered by the rapid market grow: *And when undertaking
market research consumers provided many conflicting responses in terms product &
service innovation, hence a lot of trial and error research which did not lead to sustainable product innovation – a clear example of this is the underdevelopment of entertainment sector, an important component of IR's in Las Vegas.*

Customer Innovation: To Being Flexible and Available

Being a service, tourism is a sector that benefits from the customers involvement in
the innovation process that occurs through the constant interactions that take place
with the service providers, the tourism and hospitality companies (Hjalager, 2010).
This critical interaction was noted by one of the managers interviewed (KI2):

> *We started to build and provide customers with mobile apps for them to see their performance;
> and for our staff to understand customers' behavior and to push promotions. (. . .) We maintain
> real time communications with the customers: personalized one-on-one communications (apps
> for them to communicate directly with us; they can call or text us). We will sort it out from their
> home to the airport here. In the airport, they can see everything (booking, itinerary, etc.). As a
> host we have all the information (we have all the information of guests' likes, dislikes, etc.).*

While another interviewed manager (KI4) shared: *In the past, the guests' profiles
was not streamlined. However, with an integrated CRM (Customer Relations Management) system which is outsourced, our staff know our guests' "win/loss" situation. For
those customers whose loss is greater than a certain threshold, we will intervene with
some "sweeteners".*

Innovations amongst hospitality services providers with tourists have many
manifestations such as creative tourism which has been defined as "travel directed
toward an engaged and authentic experience, with participative learning in the

arts, heritage, or special character of a place, and it provides a connection with those who reside in this place and create this living culture." (UNESCO, 2006, p. 3). While Macau has a number of tangible heritage sites which showcases its historical culture for particular type of tourists, its main attraction is being the unique gaming destination for the entire China. Regarding the gaming services offered in Macau, the manager interviewed (KI3) shared that:

> *Whereas Chinese focused customers are influenced by the characters to them, by analyzing the math. It is intrinsically part of the Chinese culture that you analyze the math by trying to find luck is about probability. So we transformed the gaming machines for this market. (. . .) And we made adjustments that has nothing to do with game rules but has to do with understanding how the customer's play. Those are the types of innovations that we have applied and are now being adopted by other gaming markets.*

Indeed as pointed out by Williams and Shaw (2011, p. 43) 'International tourists will have different experiences – amongst themselves as well as in relation to domestic tourists – while motivations, expectations and behavior are all deeply culturally imbued. However, firms can learn not only about how to innovate in relation to this market segment, but also transfer ideas to their provision of services to the domestic market.' The dynamism of the segmented market was also observed by one of the interviewed managers (KI3):

> *We build our products to service the segmentation of our customer base. We define very articulately in the way we say who are our customers, what do they need, and how do we deliver on that. And if you are really clever in the competitive marketplace, you can get this symbiotic cohabitation because everybody has strengths in certain parts of the market and they tend to gravitate to those different parts.*

Self-service technologies have been facilitated through innovations in social media providing flight information, checking-in systems, and app enabled pre-flight and post-flight purchases that gave customers greater control and flexibility on their various choices. For example, in-flight dietary requirements and seat choices, and post-flight reservations for transport and accommodation. On the use of self-service technologies, KI4 noted:

> *Using self-service technology, guests scan a QR code and through a menu are able to interact with us on their various* service *requirements and choices. We also use the QR code for the payment of bills by our guests as customers prefer to make cashless payments for retail and membership payments.*

However, regarding the use of the latest technologies for the hospitality and tourism industry, Law, Ye, and Chan (2022, p. 630) cautions that ". . . adopting smart technologies without fully understanding tourists' demands may not achieve desirable outcomes." Thus, even with all the latest technologies used to make the services innovative (e.g., AI, VR, Robots, etc.), the personal touch in hospitality needs to be present somehow in the service provision-consumption experience factor for the Chinese tourists. Thus, it was highlighted by one of the interviewed managers (KI3):

> *If companies assume that technology can be used to replace human service interactions then they are missing what should be their most important differentiator . . . . Their service touch! Technology is great at doing things but person to person interactions is the only way to truly personalize a customer experience. It's easy to mail a birthday card but having it hand delivered is so much more impactful & sustaining.*

Noting the importance of transportations role for tourism, Page, Yeoman, Connell, and Greenwood (2010, p. 120) recommend based on scenario planning that ". . . tourism may be a key revenue generator for income and taxation, but it also has reinvestment requirements to keep it competitive, accessible to consumers to make it attractive and easy to use once you arrive in the destination."

Innovativeness is also manifested by tourism companies by their strategic awareness and targeting of particular customer groups (Hjalager, 2010). The marketing strategy must be made clear on the preferred tourists that are best for enhancing the brand of the tourism destination. As advised for Macau's tourism and hospitality sector, Greenwood & Dwyer (2017, p. 596) suggest that: "Instead of asking: 'what tourists do we want to target in promotion', Macau could ask 'what tourists do we want to attract'? These are different questions. The latter involves creating the type of destination that 'ideal tourists' will wish to visit – attracting the right type of tourist is more important than attracting large numbers." As one of the managers (KI3) observed about the mainland Chinese tourists coming to Macau,

> *The Chinese customer is on a high speed journey to sophistication.* And KI3 reiterated the point of this expanding market for *Macau, Now, that concept requires rapid and transformational innovation if you are to keep pace with the rapid development of the consumer economy in China.*

Manager (KI4) pointed out that:

> *With reference to customer innovations, in the past there was only one-tier membership cards in Las Vegas. Five years after, there are now four-tier cards and even five-tier cards when you include the mass market. City of Dreams (COD) even has six-tier cards for the segmentation of their market.*

Observing the key role of organizational flexibility to enhance competitiveness underpinned by innovation in the tourism and hospitality industry, Anning-Dorson and Nyamekye (2020, p.618) noted: 'Hospitality firms must build flexible teams and networks that understand the dictates of the market and respond in real-time. One of the reasons why innovations fail in this sector is the rapid rate of change in customer needs. Building flexible organization will enhance customer needs anticipation of which innovations can address to create competitive advantage.' A vision of one of the managers interviewed (KI2) reflects this important consideration on being available to customers:

The vision I have as a future IR, it's from the moment you plan to come over to until the time you leave, everything is in one mobile app. You do all your reservations from that one mobile app, you see all the itinerary, I can push you real time promotions, and everything you spent depending of course on the level of spending you have, we can give you credit or on credit.

The process of checking in and out for hotel guests has also been speeded up:

In the past, most IRs would use the "hotel approach" for checking-in their guests. However, with the increase in the number of guests to their accommodations, the deposit was waived and the waiting time was shortened from 15 minutes to five minutes. To check-out is even faster with the guest merely having to drop their room key only. (KI4)

Product Innovation: Risks, Rewards, and the Race Against Time Amidst the Macau Gaming Boom

Let's go back to Innovation. So, I actually see the nature of innovation as steps, or evolutionary steps within the transformation of Macau. (KI3)

Product innovation comes in the forms of radical product innovation and incremental product innovation (Greeven & Yip, 2019), with frequent emphasis that there is comparatively more resistance to user adoptions for radical innovations than the latter (Eroğlu, 2019). Verganti (2009) highlighted that breakthrough technology pushes radical innovation to achieve quantum leaps in performance, whereas incremental innovation is fueled by the market through analyzing users' needs of current products to improve products and introduce solutions. Naturally, there are more incremental innovations rather than radical, and acceptance that follows, as such concepts involve slight differentiation and changes to products and services that emerge from them.

Long before the transition of power back to China, gambling has been permitted in Macau since the 1850s, and remains to be the only part of China that allows casino gambling after the handover in 1999. Bordering the South China Sea with access to a large population base, and given the Far Eastern residents' penchant for gambling, Macau was able to skyrocket to becoming a leader in the gaming market in a short period of time (Dense, 2011). When Edmund Ho Hau-wah was appointed as the first Chief Executive, he found himself in a peculiar position to transform and diversify the former Portuguese colony's economy. At the time, the gaming industry was a monopoly, and *the bottom line is: monopolies are neither innovative, progressive, nor particularly entrepreneurial* (KI3). The liberalization of the gaming sector in 2002 put an end to the four-decade sole-control of tycoon Stanley Ho and the gaming market's monopolistic era. The PRC opened up the market to foreign investors, which allowed Ho to bring in international brands including some of the world's largest gambling operators, Steven Wynn of Wynn Resorts and Sheldon Adelson of Las Vegas Sands (LVS). The growth of Macau's casinos and growth of the gaming market that followed was explosive and unprecedented, placing Macau on the global map as a transformed destination for gaming and luxury. By doing so, Ho ameliorated the negative brand proposition for Macau.

> And that's always the issue with innovation isn't it – it's a risk-reward. How far out do you want to go? So, transform and innovate the brand, and diversify the brand, transform the product, transform the quality of the experience but is always in anticipation of the consumer demand. (KI3)

Greeven and Yip (2019) noted that most Chinese companies began as imitators due to its catch-up situation, whereas Western companies were at the frontier and are highly innovative. The introduction of Venetian Macau by LVS was new to the Macau market, but it is a replica of the Venetian in Las Vegas, only bigger. Verganti (2009) introduced a third product innovation strategy – design driven innovation; which is the radical innovation of meaning. Effective design driven radical innovation characterized by the focus on socio-cultural changes over current market trends, and is a key factor of success in many products and firms worldwide (Verganti, 2009). The opening of Wynn signified the true beginning of Macau with a new image of luxury and class. The vision of Macau becoming a luxury destination because *for the Chinese market, luxury is critical. And that was going to be critical going forward as this would provide Macau a regional competitive advantage.* (KI3).

> The first innovation came in the rebranding of Macau, or rather, from the time of the freeing of Macau gaming sector and subsequent transformation from a monopoly to a competitive unit structure is a huge innovation in and of itself. (KI3)

However the opening up of Macau to the China consumer market demand grew rapidly, with resulted in the gaming sector boom experienced in Macau. *The concessionaires' responded to this demand by building capacity which limited the opportunity to diversify. There was less time to evolve and innovate in fear of lagging behind market growth and not delivering the returns expected by the investors (KI3).* As a result, the industry is identified by a level of homogeneity with slight differentiating characteristics. Although both Las Vegas and Macau share similarities in distinctive feature such as upscale shopping opportunities, fine dining, high-end fashion retail, and niche art collections, the Macau gaming industry still retains, despite efforts to diversify and evolve into the Las Vegas casino model – very much its unique gaming-oriented focus in comparison with Las Vegas (Siu & He, 2014). To achieve the Las Vegas model, not only would the industry have to diversify, but the customer segment and market would have to follow as well.

> Macau as a city, there is not much to do besides gaming. I would say 80% of those who come to Macau are here to gamble, do some shopping and eat, and whatsoever. Unfortunately, gaming is still our main focus. (KI2)

Technology Innovation: Response to Scale

And that's where you have to scale up and – you need technology. You need much more robust systems which support higher customer volumes while retaining the quality of the goods and services provided. (KI3)

Greeven and Yip (2019) noted the importance of not only pursuing technology upgrades in order to maintain competitiveness, gain technological edge, and stay ahead of copycats. In an industry in which technology-wise, there is not much room for innovation as technological products are created by third-parties and regulated. During the boom of the gaming market, more robust systems and advanced technologies were needed to handle the growth in the market. Westerners were far more advanced when it came to technology, knowledge, and experience in utilizing such technologies. In a sense, technology innovation in Macau's gaming sector was a response to scale as well as the changing expectations and demands of the Chinese customer.

On the gaming side, the role that digitalization has played was crucial in the transformation and enrichment of gaming products (such as electronic versions of the standard table games and gaming machines) as well as services offered. Aside from the implementation of new gaming systems, in order to enhance the customer gaming experience, operators launched mobile applications, upgraded websites as well as digital services. Now, every IR has a gaming system, but the level of automation is not the same across-the-board and the variation is partially due to regulations.

There was no system in place in the first gaming company when I joined the industry in 2005. All gaming was done manually including tables and reporting, and there was no automated process on the hotel side either; no automation at all. One of the biggest technological innovations for Macau was definitely the start of the implementation of gaming systems around 15 years ago (2007). (KI2)

Manual labor remained the dominant at most casinos in Macau until the integration of the RFID technology into table games, with MGM being the first in the market to launch the RFID smart tables in their casinos. How it works is that it involves "smart" baccarat tables and chips that are built specifically based on sensor technologies. Together with hidden cameras and facial recognition technology, casinos are able to track the betting behavior of gamers (Hong, 2019), to gather intelligence which is pushed back into a centralized database. This significantly enhanced the gaming experience by improving monitoring and recording of plays, preventing theft and fraud, and aiding in the promotion of responsible gaming (Wyld, 2008). One form of effective technology innovation was the one being done in MGM and this was shared by one of the managers interviewed (KI4):

Product system (which is called CCAS) is used by only one concessionaire – MGM. This system helps the company to nuance the player and to give a score for each one. For CCAS, it improves not just the accuracy and integrity of the game itself, it also creates real time data that allows the casino to know better the customer's behavior and to take action accordingly. Nowadays, data is the key power to make a difference. By having this data, the casino could use it for complex analytical capabilities. This accurate point system benefits both the customers and the company in knowing how they stand in real time. There is also a system for the gaming tables whereby management is able to track all the chips using RFID. This is a control system to make sure all goes well.

Due to Macao's IR industry's focus on the financial benefits brought forth by the gaming sector throughout the years during the aggressive growth of the gaming market, there is a recognized phenomenon that there is a significant gap in funding for technology or innovations on the non-gaming side as there was not much of a return of investment.

We had a lot of funding for the gaming side of the business, since 80%–90% of the revenues were from gaming. From the hotel perspective, even when we have 100% occupancy, 90% of that would be made up of gamers. If you are a gaming customer and you play enough, we give you a hotel room for free. (KI2)

There is also a shift in focus within the gaming sector; given that there is still relative ease when it comes to the collection and sharing of data among Chinese customers (Hong, 2019), companies have begun to recognize the value and importance in customer data and relationships. Across the industry, digital technology and big data analysis are utilized to not only gather data for marketing campaigns, but to push real-time, personalized communications, promotions to gamers in order to maintain customer satisfaction and loyalty. These are some of the critical components of customer relationship management (CRM), the industry has been known to be at the forefront of implementing CRM systems and using customer data to improve and manage customer relationships (Liu, Dong, and Zhu, 2021). Initially, each business group was working as silos, but now most companies in the gaming industry have developed centralized platforms and integrated these systems for smoother performance and service delivery.

For us as gaming IRs, whether we like it or not, the resorts were just a resort community for the gamers. But that is changing – since the decline of the VIP market five to six years ago – we realized that we cannot just focus on gaming and started to change. (KI2)

The extent of innovation and integration of technology depends heavily on factors such as social, economic, regulatory, and political (Caselli & Coleman, 2001), and Macau's gaming sector is no exception. Las Vegas and the Philippines have newer technology because of wider acceptance and easier regulations. For instance, despite the rising popularity of Artificial Intelligence (AI), it has great potential to be integrated into the gaming industry. AI-based technologies can not only be used in the optimization of revenue streams from gaming systems, it can also be used as a marketing tool to drive revenue through optimizing game design, customizing user interface and experience, and creating targeted marketing campaigns based on what would appeal to customers (Tottenham, 2019). However, it has not gained wide application due to concerns for ramifications it would have on privacy and data protection. There is much to be done using AI, but there are still many issues and challenges that come with use due to its nature as a disruptive technology with a wide range of influences (Chen, 2019).

The rigid public policy in Macau is a key reason behind the lag in innovation for the gaming industry. Online gambling is a topic that has also been frequently brought up, especially with the increasing use of mobile phones and online games in addition to expanding internet penetration. In 2021, the global market size of online gambling was valued at around USD 57.54 billion and is projected to grow from 2022 to 2030 at a compound annual rate of 11.7%. Europe alone dominated the online gambling market valued at USD 23.63 billion in the same year (Grand View Research, 2022). Legalizing online gambling in Macau could drive gaming revenue given the wide accessibility of the internet as well as its cost-effectiveness at the same time.

We would also embrace more technology if online gaming is allowed, because it is a huge market, including the metaverse and blockchain technology. A lot of companies in the United States and Europe welcome these new technologies with open arms and they gain a lot, but unfortunately not for us. (KI2)

Process Innovation: Fast Trial, Then Lean Method, and Sustainability

Let's go back to Innovation. So, I actually see the nature of innovation as steps, or evolutionary steps within the transformation of Macau, but that also requires a clear strategic objective in terms of what a diversified Macau would be like (KI3)

Greeven and Yip (2019) bring up their discovery of the trend of relying on fast trial and error for market testing, adjusting, and learning throughout two empirical research programs conducted. From start-ups to tech giants, market testing and validation, product development, as well as the designing of business models oftentimes occur simultaneously (Greeven and Yip, 2019), the concept is similar to that of the popular lean startup method (Ries, 2011). The Lean method to business processes is an approach originally derived from the Toyota Production System, the three fundamental principles are: delivery value according to customer definition, elimination of waste, and continuous improvement. Through early market validation, learning from experimentation in process, even with the absence of relevant technology, could open up avenues for success.

In 2013, in an effort to maximize operational efficiency, Sands China launched the Kaizen concept (Lean Management System) across the company to further explore ways to add value to the business. Managed by the Human Resources team, with the key objective of Kaizen in mind, set up the foundation to commit team members and enforce an on-going cycle of constant improvement of service (Sands China Ltd., 2013).

> *Once the (Kaizen) consultant came, after six months we hired him full time and started a small business unit for continuous business improvement. Then, we took all the executives to HK (and did an off site) for a high value stream engagement. We looked at the value streams and where they could see areas for improvement and we observed 15 different guidelines for different areas of business. Meanwhile, we had high value stream maps being done. Projects being assigned to the departments and at the same time we were training for in-house capability for process improvements (facilitators). The idea was that each department that was given this mission would do these events and then would continue to do these improvements henceforth. (KI1)*

Sands adopted experimentation with process innovation by attempting to further make enhancements in terms of productivity within the Kaizen method by exploring the Six Sigma method.

> *Unfortunately, to my understanding, (Six Sigma) is not as powerful a tool as Kaizen since Kaizen is a philosophy towards management while Six Sigma is a method. So I learned the hard way. I thought it was a step further. I realized that we were just implementing in the East what we had forgotten in the West. (KI1)*

The boom of the Macau gaming market was put to a halt in 2015 when China rolled out its anti-corruption campaign in 2014, targeting blatant shows of wealth (anti-extravagance) by public officials. Melco Resorts & Entertainment (Melco) Chairman and CEO Laurence Ho noted in a 2018 CNBC interview, that despite the negative impact that led to two tough years, they had the time to focus on the Kaizen practice and implementing it on everything from purchasing practices to labor practices (CNBC & Tan, 2018). Ho goes into further explanation during an interview the following year with Inside Asian Gaming while discussing his vision for Macau as well as Japan ventures (Cohen, 2019):

> *"Operationally, we believe in "kaizen". There's a lot of Japanese philosophies that we like, but kaizen is something we've been working on for the past three or four years. We have our own in-house trained and registered kaizen trainers. It's about continuous improvement, how we can do everything better."*

Melco continued to employ the Kaizen method throughout the years, on the sustainability front, Melco has announced the implementation of a Kaizen review of the process of waste handling at City of Dreams, Altira Macau, and Studio City resorts with the objective to improve processes by reviewing current processes, and identifying gaps in the capturing and recording of data (Melco Resorts & Entertainment, 2021).

With the COVID-19 pandemic along with the downfall of the VIP junket sector, casino operators are forced to a halt having been struck blow after blow with losses greater than the previous. Moving forward, adopting the wide use of the fast trail and learning method could be favorable to help the gaming industry navigate through turbulent waters, as experimentation and risk-taking are in alignment with Chinese pragmatism (Greeven & Yip, 2019).

Organization Innovation: Measuring Innovation In An Efficiency-driven Culture

Organizational innovations which are also described as institutional innovations (Hjalager, 2010, p. 3) is "a new, embracing collaborative/organizational structure or legal framework that efficiently redirects or enhances the business in certain fields of tourism." This institutional innovation occurred at the start of the liberalization of Macau's gambling industry which one of the managers interviewed (KI3) noted,

> With Macau's return to China its leaders understood the need to diversify Macau" And, therefore the need to innovate Macau: and this led to opening up the gaming market by bringing in competition to the monopoly in place at that time. I think they also understood that Macau required an increase in supply if it was to provide customer choice and also in terms of the number of global brands to compete with the expected regional competition.

Martínez-Román, Tamayo, Gamero, and Romero (2015) analyzed how innovativeness (i.e. through the determinants of knowledge, organization, and human factors) influenced the business performances of SMEs from the hospitality industry in Spain. SMEs belonging to the tourism industry were viewed to have limitations in investing in knowledge management systems – both within (micro) and outside (macro) of the organizations – because of their resource constraints and eventual lack of innovations (Raisi et al., 2020). Nonetheless, SMEs do play a part in having an entrepreneurial role in the development of smart tourism together with the larger and resource-rich advantaged multinational corporations (Williams, Rodriguez, & Makkonen, 2020).

Based on a study of local SMEs relationship with large Integrated Resorts during rapid economic growth period in Macau, Cuervo and Cheong (2017, p. 331) noted: "The survey in this study revealed that local SMEs are not without disadvantages relative to large MNCs in Macao. SMEs enjoyed greater flexibility and managed a quicker decision-making process relative to their larger counterparts. In particular, local SMEs have the local contacts and can collaborate quicker with their network of local stakeholders which Multinational corporations (MNCs) lack and need. Local SMEs that managed to do business with MNCs have learned to build on their experiences an international dimension and standard that bodes well for future international expansion." However, being in a small geographic jurisdiction, the SMEs in Macau were disadvantaged due to their having to compete with the large MNCs in Macau, when hiring from a limited pool of skilled human resources, and from the negative effects of increasing rents of real estate during the rapid economic periods in Macau (Cuervo & Cheong, 2017). As Cuervo and Cheong (2017, p. 331) pointed out about local SMEs vis-à-vis large foreign MNCs: "SMEs cannot match the pay scale, benefits and training provided by the MNCs, as these have the financial resources, management experiences and marketing abilities that SMEs cannot match. MNCs generally have a strong brand underpinned by their size and reputation."

From an 18-month research of a large tourism company TUI, Buitjtendijk, van Heiningen, and Duineveld (2021) observed that an understanding and interpretation of innovation was challenging for the management of a large tourism organization that has an efficiency-driven culture. Measuring the 'unit of analysis' of innovation can be vague as it is a multi-dimensional reality, and thus Buitjtendijk, van Heiningen, and Duineveld (2021, p. 9) counsel "We argue that it is precisely the anticipated and unanticipated, wanted and unwanted, reality effects that simultaneously strengthen and limit innovativeness in organisations." This was similarly experienced by one of the managers interviewed (KI1) who had the following salient observations:

> The idea of knowledge transfer was bringing to every team the same level of knowledge that they would not get – it was a leverage and knowledge transfer happening really fast. Because we were pushing also behavioral change in terms of doing the opposite of what was going on now, which is share information, using information in collaboration, and share the success. . . . The idea was to come together and have transparency in the work you do. To be able to inspect your work and to do it much more frequently. Because the outcome will change. There was no predetermined outcome of the project: that was always being shaped through the contours of the business . . . Agile was invented for complex scenarios where you cannot forecast the outcome but you can understand the connections and how they play and you can find direction thru how agents in that complex system interact. The important thing are the connections – how we interact at work – and not necessarily the outcomes that you and I need to produce.

The transparency and sharing of information for an effective knowledge management system was also shared by KI4:

> We have a finance/Human Resources management platform for internal use for us to know the work days / leaves of all staff. This transparent platform is also used for knowing the P&L of each business unit in our company.

The challenges toward measuring and interpreting innovation does not mean to neglect the efforts at sustaining innovations in tourism companies. For example, Wu (2020) observed from an empirical study of 217 Chinese tourism companies that through the quality of their human capital and absorptive capacity of specialized knowledge, coupled with the mediating role of asset specificity management, these Chinese tourism companies were able to achieve excellent innovation performance. Thus, Wu (2020, p. 924) suggested that 'to improve tourism innovation performance, organizations should insist on continuous investment in sustainable human capital without hesitation for a long time. In this case, managers can design and implement specially designed and unique employee training schemes, which can contribute to improve the knowledge and experience as well as promote the progress of new skills and ideas of staff, achieving better innovation performance.' Unfortunately, such investments on innovations may not be understood or appreciated by top management. For example, one of the interviewed managers (KI2) shared his experiences:

> There are projects that I am still talking in the past seven years. He believes we need to do it but the investment is too big. The owner looks more about security (protection of the investment; protect the "house") rather than to increase customers' experience (which is not a priority). Plus, it all depends on how the economy is doing. Here, just one person makes the decision.

When asked the question, 'Innovation is for the long term (5 years or more) and oftentimes, when you look at the ROIs you tend to look at the short term returns. Is this an issue you encountered and what can be done?' The manager's (KI2) reply was:

> Well it depends on each company. I worked in a company where the technology department worked under the CFO and now I report directly to the owner and not the CFO. It really depends. If the CFO sees there is return but the owner will see that even he believes there is a return he may not invest because the owner is taking his money. I will be taking away my money from something else. He did not want to spend money. However, in the past five years, he started to think more on technology. The owner now understands (in the past five years) we need to embrace technology and to innovate and change as everybody else is doing.

Interestingly one manager (KI4) noted:

> If we are to assess innovations according to their return on investments (ROI), five years is considered an acceptable time frame.

"Innovations" in Macau's IR Company's Annual Reports & Macau's Tourism Master Plan

The annual reports from the past ten years (2012–2021) of the six companies managing Macau's Integrated Resorts were analyzed by identifying the number of times the word "innovate" (or related words, "innovative", "innovation", etc.) appeared in these reports. While this identification is only indicative of the management's conscious efforts at communicating to key stakeholders about their efforts toward innovations, it nonetheless is a measure of intent worth observing and reporting. The identified portions of the annual reports with the search word "innov-" were extracted for further analyzes. Starting from 2012, to 2013, and eventually to 2021, the extracted parts were classified as a "relevant keyword" when this appeared for the first time in the years analyzed. Furthermore, these "relevant keywords" would be counted only one time to avoid double-counting when the concept is repeated elsewhere in the same year's annual report or in later years. Based on this exercise, the following are the total number of times the "relevant keywords" appeared during the ten year period from 2012 to 2021: SJM (4 times), Wynn (5 times), GEG (15 times), Sands (19 times), Melco (41 times), and MGM (42 times).

For example we note from the SJM Holdings Limited Annual Report 2016 the "relevant keywords" which are presented in **bold** format in the following sentences (SJM Holdings Limited, 2022, p. 91 of Annual Report 2016): 'SJM is a supporting organization for the Fifth IFCE [International Forum for Clean Energy] held in Macau on 29 and 30 November 2016, the theme of which is "Clean and Low-carbon Efficient and Energy-saving'. It focuses on four reforming and development action roadmaps for pragmatic discussion including energy production and consumption, technological **innovation**, institutional mechanisms and international cooperation. Dr. So officiated the opening ceremony, made an opening speech and received the Certificate of 'Carbon Neutral Conference' awarded by the China Green Carbon Foundation on behalf of all attendees at the Fifth IFCE. A number of SJM's senior executives joined the Fifth IFCE and attended the programmes and activities it had organised.'

Two year after, the SJM Holdings Limited Annual Report 2018 noted (SJM Holdings Limited, 2022, p. 19 of Annual Report 2018): 'SJM participated in activities during "Macau Energy Saving Week", including the One-hour Lights Off and Dress Casual Summer campaigns. The Company is co-organiser of the Seventh International Forum on Clean Energy. The forum, held in Macau in December 2018, brought together representatives from domestic and foreign governments, ambassadors, as well as delegates from industry associations, enterprises, universities and research institutions to present, discuss, and exchange views on the clean and efficient use of hydropower, wind power, photovoltaics, nuclear energy, electricity and coal, as well as smart energy and energy block-chain technology, focusing on promoting industrialisation and international cooperation for clean energy and smart energy technology. The forum also witnessed the launch of the 2018 Blue Book on International Clean Energy Industry Development and the 2018 Blue Brook on Smart Energy Industry **Innovation** and Development.'

The two other "relevant keywords" for SJM appeared in the "Chairman's Statement" found in the SJM Holdings Limited Annual Report 2021 (SJM Holdings Limited, 2022, p. 3 of Annual Report 2021): 'Our mission is to build a long-term sustainable presence, focusing primarily on Macau, **innovative** in developing and adapting our businesses, environmentally responsible and concerned with the well-being of our customers, patrons and the local community . . . The **innovative** nature that we have shown in the past is displayed by our partnerships with the fashion houses of Versace and Karl Lagerfeld in the design of two of the hotel towers on the property.'

GEG reported in 2012 (Galaxy Entertainment Group Limited, 2022, pp. 12–13 of Annual Report 2012): 'The Group's continued success derives from being a local operator with an intrinsic understanding of Asian customer tastes and preferences. Our 'World Class, Asian Heart' philosophy permeates all aspects of the business, from our spectacular properties, to our **innovative** products, services and people.'

And for Melco, we note the Chairman and CEO's statement found in the 2012 Annual Report of Melco, (Melco International Development Limited, 2022, p. 12 of Annual Report 2012): 'Our **innovative** ideas for creating unique integrated resorts and exciting never-seen-before attractions in City of Dreams continued to attract visitors. In just over two years since its debut, The House of Dancing Water has welcomed nearly two million spectators and has truly become one of Macau's and Asia's entertainment landmarks.'

From the Wynn Macau 2015 Annual Report (Wynn Macau, Limited, n.d., p. 48 of Wynn Macau 2015 Annual Report) we observe early efforts toward innovation in the learning and advancement of their people: 'In 2007, we established the Wynn Academy, which offers **innovative** and tailor-made programs to assess and harness the potential growth of our employees.'

The MGM China Holdings Limited 2012 Annual Report highlighted (MGM China Holdings Limited, n.d., p. 20 of 2012 Annual Report): 'We continue to introduce new **innovative** gaming products to enhance customer experience. We provide regular professional and service training to our employees with the goal of building a culture of execution excellence. The investment we made in our products and our employees was an indispensable factor that allowed us to achieve the continuous growth and financial results in 2012.'

The sample of "innovations" as reported in the annual reports point to unique innovative strategies advocated by the various IRs in Macau: some focusing more on product innovations, whereas others on technology innovations, and some still on business model innovations with emphasis on people-related learning. However, Narduzzo and Volo (2018, p.739) rightly emphasized that 'In view of tourism's multifaceted characteristics and its relational nature, its innovation should be framed within a system of interdependencies.' What is essential is for the IRs to incorporate the learning process within the innovations that were implemented and which should have positive benefits for the enterprise, and its stakeholders – including the Special Administrative Region of Macau in China.

One key stakeholder is the Macao Government Tourism Office (MGTO) which has recently been tracking the implementation of various action plans toward being a "World Centre of Tourism and Leisure" (MGTO, 2021). Based on their research using network analyzes, Raisi, Baggio, Barratt-Pugh, and Wilson (2020) highlighted the critical role of inter-organizational knowledge transfers for the basis of innovations in tourism destinations. This is in line with one of the eight key objectives of the MGTO's Master Plan: "Deploy and Utilize Innovative Technology". To improve the travel experience of tourists in Macau with the support of innovative technologies, there are three phased plans (MGTO, 2021): (1) Continue to promote innovative technologies to enhance visitor experience. For example, through AR/VR, tourists will experience enhanced interactive display of information on tourist attractions and use AI narration and interactive programs to enrich their experience. (2) Improve the use of network and electronic supporting facilities, and expand their coverage. (3) Assist SMEs to adopt innovative technologies. For example, SMEs are encouraged to adopt the zero-contact business model by applying more technology for innovative processes in marketing, sales, and payment systems, so as to enhance their productivity, and competeveness. Such over-arching efforts by the MGTO toward innovations for the tourism, leisure, and hospitality industry will definitely benefit the IRs of Macau that are pursuing their own enhanced innovations for adequately diversifying Macau's economy.

Driving Forces & Questions for Future Scenario Planning

Rather than concluding and making concrete recommendations about innovations for Macau's IRs, this study seeks to identify driving forces and to raise questions that may shape the future operations and performance of IRs.

Macau is at a crossroads with the government's recent open tender invitation with bidders expected to submit their proposals from July 29 to September 14, 2022 (Macau Business, 2022). The bidder's proposals will be evaluated according to seven criteria by a committee that will decide on the eventual awarding of the future six gaming concessionaires for a period of ten years. The seven criteria are: (1) the amount of the premiums the bidders propose; (2) their plan in exploring non-Chinese markets; (3) their experience in casino gaming and related fields; (4) the benefits of their gaming and non-gaming investments; (5) their casino management plan; (6) their proposal of overseeing and preventing illicit activities in casinos; and (7) social responsibilities.

The seven criteria provides ample opportunities for budding concessionaires to present their detailed plans that incorporates specific dimensions of feasible innovations. Moreover, these innovations should support the adequate diversification which IRs will contribute toward Macau's being a center for tourism, hospitality, and leisure.

A few critical driving forces and questions that may shape the future scenario planning to be done by budding concessionaires include: (1) On the use of an approved government e-currency, "What is the probability of the introduction of a government mandate that some gambling-related transactions must use an approved government managed digital RMB in Macau?" (2) Assuming that the government's managed digital RMB is approved for use in online gambling transactions, "What is the probability of approving the promotion of online gaming to be operated by the Macau-based gaming concessionaires?" (3) With the benefits of increased market competition in the region, "What is the prospect of the opening of another approved and legal gambling destination in a city or region in China?" (4) With efforts at integrating Macau further into the Greater Bay Area by jointly managing non-gaming related tourists' attractions in Hengqin island, "What are the fiscal benefits that would incentivize Macau's integrated resorts to invest in world-class tourists attractions in Hengqin island?" (5) Any promotion of innovations for tourism, gaming, and hospitality that were organically developed by Macau-based SMEs and IRs which are recognized by the World Intellectual Property Organization (WIPO) as an intellectual property right should be amply rewarded. "What attractive funding and rewards would develop commercially viable innovations (i.e. intellectual property rights) that are directed toward the tourism, gaming, and hospitality industry?" (6) There have been some very successful multinational corporations that are from small open economies (e.g. Ivoclar Vivadent and Hilti are both from Liechtenstein) that could serve as an exemplary example for supporting international entrepreneurship in Macau. That begs the question, "How could the IRs in Macau be incentivized to contribute toward investing and supporting local entrepreneur and businesses that have good potential to develop into successful multinational corporations?" (7) The development of foreign markets with the underlying expansion of non-gaming elements is the key focus for casino concessions, given that the current non-gaming features of Macau's IRs have questionable returns, "would the Las Vegas model be the favorable direction moving forward? If not, what ways could the IRs diversify and optimize "non-gaming projects" to successfully expand and diversify the currently gaming-driven visitors?" (8) The IR industry is in the midst of a historically bad three years due intermittent lockdowns and travel restrictions caused by the COVID-19 pandemic, with Macau maintaining China's "zero-Covid" policy while many other countries in Asia have adopted the "Living with COVID" policy in recent months. "How will the IRs navigate this road to recovery that is entirely dependent on non-residents being permitted to freely enter in and out of Macau?"

At this watershed period for Macau's Integrated Resorts, a correct understanding of the common scenario planning methods available (Fergani, 2020) would be of assistance for management in preparing for their strategies and implementation action plans.

Discussions

This study suggests that vulnerable companies are those that have neglected to innovate in and to learn from their business operations even when they were doing well during a period of a privileged position in the market. This contrast with sustainable companies that have wisely invested in learning from innovations to constantly improve their diversified businesses. Our study found that the integrated resorts in Macau have manifested some aspects of innovations but with ample room for further development to effect sustainable diversification from a heavy dependence on gaming revenues.

The mainland Chinese tourists who are Macau's main visitors were characterized as a market segment which is on a high speed journey of sophistication needing special attention to learn from their behavior and adopt to their needs (Williams and Shaw, 2011; Anning-Dorson and Nyamekye, 2020). While the use of CRM systems, self-service technologies, and mobile apps has been employed to enhance the service experience of visitors to Macau (Liu, Dong, and Zhu, 2021), there is no substitute for the personalized service touch in the hospitality industry as pointed out by some interviewees and as highlighted by Law, Ye, and Chan (2022).

With increased volume and sophistication of customers, Macau's integrated resorts had to scale up through technology innovations (Greeven and Yip, 2019) and AI-based technologies (Tottenham, 2019). One of our interviewees highlighted the switch over past years when leaders in the industry began to embrace technology and see it as a necessary tool for the advancement and sustainability of business. For example, the implementation of smart tables and RFID in gambling chips, was introduced by one of Macau's integrated resorts to improve the gaming experience and to better track player behavior throughout playtime (Hong, 2019). With the use of AI amongst other forms of technology, there are privacy concerns due to data-sourcing technologies still needs to be addressed (Chen, 2019). Moreover, the strict regulations that Macau's gaming operators face hinder expansion of the technology aspect of innovations within the industry. For example, the large appetite of online gaming unfortunately cannot be tapped into by Macau's IR operators due to the government being against online-gambling as mentioned by one of our interviewees.

Our research observed that the re-branding of Macau was given a boost as casino concessionaires opened attractive high-end luxury real estate development which signified a form of innovative design (Verganti, 2009). This was a major break from the past which was described by one manager interviewed in our study that: 'monopolies are neither innovative, progressive, nor particularly entrepreneurial.' From a monopoly to a privileged "competitive" market, the six concessionaires were not adequately diversified even after twenty years of operations. This was aptly described by one key informant to the study who pointed out: "unfortunately, gaming is still our main focus" and therefore Macau is still very far from the successfully diversified "Las Vegas" model (Siu & He, 2014). Our study also learned that the reason for not diversifying from and innovating in the gaming business was because the concessionaires were distracted with building capacity to meet the growing demand for gaming. Not investing in the *same* business would have meant losing market share and reducing profits to their shareholders' dissatisfaction.

Process innovations were manifested in various integrated resorts in Macau as they employed lean management and continuous improvements (Ries, 2011). Process innovations manifested through a step-by-step evolutionary approach rather than radical innovations (Greeven & Yip, 2019) bodes well for the operations of the integrated resorts in Macau. Much of the connections between innovations with learning occur during process innovations which are necessary for sustainable companies.

The managerial innovations which are underpinned by people, operational efficiencies and flexibility (Hjalager, 2010) was also observed by the managers interviewed in this study. For example, recruiting high quality locals with globally competitive skills, attitudes, and experiences has bolstered the Macau brand which aligns well with the suggestions for productivity enhancement in the tourism industry (Brien, Vidwans, & Dutt, 2022). Moreover, reducing overhead and costs control measures were also clearly articulated by the interviewees as part of the integrated resorts' business model innovations being undertaken in Macau. As to the need to be flexible and adapt quickly to the changing market demand (Anning-Dorson & Nyamekye, 2020), unfortunately, one manager observed that effective application of market research done in Macau was hampered to serving the rapid market growth. Consequently, they failed to deliver on sustainable innovative products, such as in developing the entertainment sector, similar to that in well-established integrated resorts in Las Vegas. However, from our interviews, we ought to acknowledge that the "Las Vegas" model, though attempts were made to push its implementation to diversify the industry, that it is due to demand of the market's main customer base being vastly different to that of Las Vegas. As pointed out that the majority of Macau's visitors arrive to the gaming hub with the main goal of gambling, things such as entertainment and art are not actively sought out. Even hotels and restaurants at Macau's integrated resorts seem to be meant as added features to the casinos with the intent to prolong the stay of visitors at their properties.

Finally, with reference to organizational innovations, given the complexity of knowledge sharing and the multi-dimensional nature of innovations (Buitjtendijk, van Heiningen, & Duineveld, 2021), an interviewed manager observed that there was some effort toward greater transparency, cross-functional collaboration, and relevant training in Macau's integrated resorts for effective learning and innovations to take place (Wu, 2020).

Conclusion

This paper investigated the extent of innovations amongst the integrated resorts in Macau over the past two decades based on the six paths of innovation (Greeven & Yip, 2019) within an overarching framework that analyzes companies as being within the range of sustainable or vulnerable depending on the extent of innovation and learning. Some of the more salient findings is that more efforts are needed toward developing product, managerial, and organizational innovations in the integrated resorts in Macau. For example, it was unfortunate that the leisure beyond-gambling aspects of these tourism destinations did not develop a more attractive and diverse entertainment offering which could have catered to the large and sustainable demand from Chinese customers. Likewise, greater flexibility as informed by correct and swift market research would have strengthened the business model innovations needed by management of these integrated resort companies to cater to emerging trends in the market.

Further studies could focus on a number of research areas such as on how integrated resorts in Macau could diversify its revenues sources by internationalizing outside China. Yet, other research may focus on the feasibility of building up capacity in creating top class non-gaming leisure and hospitality attractions in Hengqin Island as an attractive tourism hub in the Greater Bay Area for the entire China. To support the tourism hub, further research could investigate the scenario of having tertiary educational institutions from all over China setting up world-class learning facilities in Hengqin Island for their students to learn and specialize in hospitality, gastronomy, leisure, and tourism. What are the benefits from these diverse students' learning-and-innovating together to support the internationally competitive joint tourism destination of Macau-Hengqin Island?

This study is not without limitations. For example, the limited number of key informants interviewed may have their own perceptions about the meaning and differences between innovation and diversification. Thus, future research could compliment this study by investigating the topic of innovation and diversification of integrated resorts in Macau through a more rigorous theoretical framework and developed hypotheses to be tested through a much larger sample of respondents.

References

Anning-Dorson, T., & Nyamekye, M. B. (2020). Be flexible: turning innovativeness into competitive advantage in hospitality firms. *International Journal of Contemporary Hospitality Management*, 32(2), 605–624.

Brien, A., Vidwans, M., & Dutt, C. S. (2022). The tourism productivity challenge: are we measuring the right factors, and is productivity growth unlimited? *Current Issues in Tourism*, pp. 1–14; published online 20 February 2022; https://doi.org/10.1080/13683500.2022.2038091

Buitjtendijk, H., van Heiningen, J. and Duineveld, M. (2021). The productive role of innovation in a large tourism organization (TUI). *Tourism Management*, 85, Article 104312, pp. 1–11; published online 2 March 2021; https://doi.org/10.1016/j.tourman.2021.104312

Cao, A. Shi, F. and Bai, B. (2022). A comparative review of hospitality and tourism innovation research in academic and trade journals, *International Journal of Contemporary Hospitality Management*, 34(10): 3790–3813. https://doi.org/10.1108/IJCHM-11-2021-1443

Caselli, F., & Coleman, W. J. (2001). Cross-country technology diffusion: The case of computers. *American Economic Review*, 91(2), 328–335

Chen, H. (2019). Success factors impacting artificial intelligence adoption: Perspective from the telecom industry in China (Doctoral dissertation, Old Dominion University). Retrieved from https://digitalcommons.odu.edu/businessadministration_etds/102

CNBC & Tan, C. (2018). CNBC Transcript: Lawrence Ho, Chairman and CEO, Melco Resorts & Entertainment. https://www.cnbc.com/2018/05/30/cnbc-transcript-lawrence-ho-chairman-and-ceo-melco-resorts-entertainment.html

Cohen, M. (2019). Going Global. Inside Asian Gaming. June 29. Retrieved on 3 August 2022; https://www.asgam.com/index.php/2019/06/29/going-global/

Cuervo, J. C. & Cheong, K. U. (2017). Protecting the survival of local SMEs during rapid tourism growth: ongoing lessons from Macao. *Worldwide Hospitality and Tourism Themes*. 9(3), 316–334,

Dense, J. (2011). Gambling and Globalization: Still a Bad Bet. *Gaming Law Review and Economics*, January, Volume 15, Issue 1–2, pp. 17–25. https://doi.org/10.1089/glre.2011.15103

Dess, G. G., McNamara, G., & Eisner, A. B. (2016). *Strategic Management: Text & Cases*, 8th Edition, McGraw-Hill: New York.

DSEC (2019). Analysis Report of Statistical Indicator System for Moderate Economic Diversification of Macao 2018 December 13, http://www.dsec.gov.mo/Statistic/General/SIED/2018.aspx? lang=en-US; retrieved 26 July 2022

Eroğlu, I. (2019). Effects of Innovation types on product identities: does radical innovation lead to a more integrated product identity? *International Journal of Innovation*, 7(2), 252–276

Fergani, A. (2020). "4 Archetypes, Shell, 2x2: Top Three Scenario Planning Methods Explained and Compared", June 26, Predict Newsletter, retrieved on 9 August 2022 from https://medium. com/predict/4-archetypes-shell-2x2-three-scenario-planning-methods-explained-and-compared-d2e41c474a37

Galaxy Entertainment Group Limited (2022). Annual Report of various years; https://www.galaxyen tertainment.com/en/investor/financial-reports

Grand View Research (2022). GVR Report cover Online Gambling Market Size, Share & Trends Analysis Report By Type (Sports Betting, Casinos, Poker, Bingo), By Device (Desktop, Mobile), By Region (North America, Europe, APAC, Latin America, MEA), And Segment Forecasts, 2022–2030. Report ID: GVR-3-68038-474-1. Report overview: https://www.grandviewre search.com/industry-analysis/online-gambling-market. Accessed July 31, 2022.

Greenwood, V. A., & Dwyer, L. (2017). Reinventing Macau tourism: gambling on creativity? *Current Issues in Tourism*, 20(6), 580–602.

Greeven, M. J., and Yip, G. S. (2021). Six paths to Chinese company innovation. *Asia Pacific Journal of Management* 38, 17–33, Published online 6 June 2019; https://doi.org/10.1007/s10490-018-9635-3

Hjalager, A-M. (2010). A review of innovation research in tourism. *Tourism Management*, 31, 1–12.

Hong, J. (2019). China's big brother casinos can spot who's most likely to lose big. Bloomberg Wire Service. June 25. Retrieved from https://www.bnnbloomberg.ca/china-s-bigbrother-casinos-can-spot-who-s-most-likely-to-lose-big-1.1278496

Law, R., Ye, H., & Chan, I. (2022). A critical review of smart hospitality and tourism research. *International Journal of Contemporary Hospitality Management*, 34(2), 623–641.

Liu, M.T., Dong, S., & Zhu, M. (2021). "The application of digital technology in gambling industry", *Asia Pacific Journal of Marketing and Logistics*, 3(7), 1685–1705.

Liu, C., & Lin, Y. (2022). Macau's sustainability and diversification. *Business Economics*. Published online on 4 June. DOI: https://doi.org/10.1057/s11369-022-00260-9

Macau Business (2022). BREAKING NEWS – Gov't launches public tender to award gaming concessions, July 28. Retrieved on 8 August 2022; https://www.macaubusiness.com/break ing-news-govt-launches-public-tender-to-award-gaming-concessions/

Martínez-Román, J. A., Tamayo, J. A., Gamero, J., & Romero, J. E. (2015). Innovativeness and business performances in tourism SMEs. *Annals of Tourism Research*, 54, 118–135,

Melco International Development Limited (2022). Investor Relations: Financial Reports. Melco International Development Limited Annual Report of various years. https://www.melco-group. com/en/Reports.html.

Melco Resorts & Entertainment Sustainability Report (2021). https://www.melco-resorts.com/sus tainability/doc/Melco_SustainabilityReport_2021.pdf

MGM China Holdings Limited (n.d.). Annual and Interim Reports, various years; https://en.mgmchi naholdings.com/IR-Annual-and-Interim-Reports

MGTO (2021). Macao Tourism Industry Development Master Plan Review Report and List of Action Plan, November, https://masterplan.macaotourism.gov.mo/2021/index_en.html; retrieved on 3 August 2022.

MSAR (2007). "Policy Address -Translated Copy – for the Fiscal Year 2007 of the Macao Special Administrative Region (MSAR) of the People's Republic of China", https://www.gov.mo/en/content/policy-address/year-2007/; retrieved 25 July 2022.

MSAR (2019). "Policy Address -Translated Copy – for the Fiscal Year 2019 of the Macao Special Administrative Region (MSAR) of the People's Republic of China"; retrieved 25 July 2022; https://www.gov.mo/en/wp-content/uploads/sites/2/2018/12/en2019_policy.pdf

MSAR (2020). "Policy Address -Translated Copy – for the Fiscal Year 2020 of the Macao Special Administrative Region (MSAR) of the People's Republic of China", retrieved 25 July 2022; https://www.gov.mo/en/content/policy-address/year-2020/

MSAR (2021). "Macao SAR Government publishes second Five-Year Plan", Government Portal of Macao Special Administrative Region (MSAR) of the People's Republic of China; Government Information Bureau; published on 16 December 2021; https://www.gov.mo/en/news/248968/; retrieved 26 July 2022.

Narduzzo, A., & Volo, S. (2018). Tourism innovation: when interdependencies matter. *Current Issues in Tourism*, 21(7), 735–741.

Page, S. J., Yeoman I., Connell, J., & Greenwood, C. (2010). Scenario planning as a tool to understand uncertainty in tourism: the example of transport and tourism in Scotland in 2025. *Current Issues in Tourism*, 13(2), 99–137.

Raisi, H., Baggio, R., Barratt-Pugh, L., & Wilson, G. (2020). A network perspective of knowledge transfer in tourism. *Annals of Tourism Research*, 80, Article 102817, pp. 1–13; available online 30 October 2019; https://doi.org/10.1016/j.annals.2019.102817

Ries, E. (2011). *The Lean Startup: How Today's Entrepreneurs Use Continuous Innovation to Create Radically Successful Businesses*. New York: Crown Business.

Sands China Ltd. Annual Report. (2013). https://investor.sandschina.com/static-files/471f94cd-0de3-4cfe-93de-9e9b5159f881

SUI, Ricardo, C. S. & Miao HE. (2014). Integration of the Macao gambling room model and the Las Vegas casino resort model. China's Macao transformed: challenge and development in the 21st century, pp. 97–117; Hong Kong: City University of Hong Kong Press; ISBN 9789629372071.

SJM Holdings Limited (2022). SJM Holdings Limited Annual Report of various years. https://www.sjmholdings.com/en/investor-relations/financial-reports.

Tottenham, A. (2019). AI and Gambling. CDC Gaming Report Inc. Retrieved from https://www.cdcgamingreports.com/commentaries/ai-and-gambling/; accessed July 31, 2022.

UNESCO (2006). Towards sustainable strategies for creative tourism: Discussion report of the planning meeting for 2008 international conference on creative tourism, CLT/CEI/CID/2008/RP/66, October 25–27, Santa Fe, New Mexico.

Verganti, R. (2009). *Design Driven Innovation: Changing the Rules of Competition by Radically Innovating What Things Mean*. Cambridge, MA: Harvard Business Press.

Williams, A. M., Rodriguez, I., & Makkonen, T. (2020). Innovation and smart destinations: Critical insights. *Annals of Tourism Research*, 83, Article 102930, pp. 1–10; available online 12 May 2020; https://doi.org/10.1016/j.annals.2020.102930

Williams, A. M., and Shaw, G. (2011). Internationalization and innovation in tourism. *Annals of Tourism Research*, 8(1), 27–51.

WIPO (2021). Global Innovation Index 2021: Tracking Innovation through the COVID-19 Crisis. Geneva: World Intellectual Property Organization (WIPO); DOI: 10.34667/tind.44315; retrieved on 28 July 2022.

WIPO (2020). WIPO Director General Daren Tang: the Greater Bay Area, a World-Class Center for Scientific and Technological Innovation. November 16, World Intellectual Property Organization (WIPO). https://www.wipo.int/about-wipo/en/offices/china/news/2020/news_0028.html; retrieved on 28 July 2022

WIPO and UNWTO. (2021). Boosting Tourism Development through Intellectual Property, Geneva: WIPO. DOI: https://doi.org/10.18111/9789284422395; retrieved on 28 July 2022.

World Travel & Tourism Council (2008). The 2008 Travel & Tourism Economic Research MACAU, London, 1 February.

Wu, A. (2020). Performance: Linking perspectives of asset specificity, intellectual capital, and absorptive capacity, *Journal of Hospitality & Tourism Research*. 44(6), 908–930.

Wyld, D.C. (2008). Radio frequency identification: Advanced intelligence for table games in casinos. *Cornell Hospitality Quarterly*. 49(2), 134–144.

Wynn Macau, Limited (n.d.), Annual & Interim Reports, various years; http://en.wynnmacaulimited.com/financial-information/annual-reports/

List of Contributors

Yumin Cao
University of Macau, Macau S.A.R.
China

Vanessa Madalena Crestejo
University of Macau, Macau S.A.R.
China

Javier Cuervo
University of Macau, Macau S.A.R.
China

Felicitas Evangelista
Western Sydney University, Penrith
NSW, Australia

Jacky Hong
University of Macau, Macau S.A.R.
China

S. H. Kong
University of Macau, Macau S.A.R.
China

Shenxue Li
University of Kent, Canterbury
United Kingdom

Lancy Mac
University of Macau, Macau S.A.R.
China

Gordon Redding
INSEAD, France

Chris Rowley
University of Oxford, Oxford
United Kingdom

Robin Snell
Hang Seng University of Hong Kong
Hong Kong S.A.R.
China

Yixin Sun
University of Macau, Macau S.A.R.
China

Xi Zhao
University of Macau, Macau S.A.R.
China

Yibing Zhang
University of Macau, Macau S.A.R.
China

https://doi.org/10.1515/9783110715002-009

List of Figures

https://doi.org/10.1515/9783110715002-010

List of Tables

https://doi.org/10.1515/9783110715002-011

Index